This Broken Planet

*[Earth's beginning and end,
and its unseen Water Layer]*

by

Gary Wentz

Contents

This Broken Planet
Second Edition

Copyright © 2025 Gary Wentz

All rights reserved.

ISBN: **9798313316338**

Published by Amazon KDP

 GaryTheGospelGuy.com

Printed in the United States of America

Unless otherwise indicated, all Scripture quotations are taken from the Holy Bible, New Living Translation, copyright © 1996, 2004. Used by permission of Tyndale House Publishers, Inc., Carol Stream, Illinois 60188. All rights reserved.

Contents

PREAMBLE 1
    WHAT THIS IS 1
    THE VISION ITSELF 2
    UNDERSTANDING THE VISION 3
    YOUNG EARTH CREATIONISM & SCIENCE 4
    WHY THIS MATTERS 5
    HOW TO TAKE THIS 5
    WHY ME? 7
    IN GOD'S IMAGE 10
    FRAGILE HEARTS AND MINDS 12

CHAPTER 1 | COMPETING WORLDVIEWS 15
    COSMOLOGY 15
    THE PAST: ABSTRACT OR REALITY? 18
    MY COSMOLOGY 20
    THE SPIRITUAL SIDE OF THINGS 22
    GOD 23
    A SPIRITUAL IMAGE OF THE CREATION 26
    BIBLICAL COSMOLOGY 31
    PRECISION 36
    MODEL PREAMBLE 37
    GROUP DISCUSSION QUESTIONS: 44
    ANSWERS: 44

CHAPTER 2 | THE BROKEN PLANET MODEL 45
    THE MANTLE 48
        Its Irreducible Complexity Reduced 48
        Heat 51
        Radiation 52
        Pressure 54
        Division 55
        Magnetism 57
        Magnetic Fluctuations in the Crust 60

Contents

- *Tectonics and Subduction* 61
- THE WATER LAYER 63
  - *Depth* 63
  - *Function* 66
  - *The Flood* 69
- THE CRUST LAYER 72
  - *Composition* 73
  - *Function* 74
  - *Surficial Water* 75
  - *Water Table* 76
  - *Other Features* 77
- PRE-FLOOD WEATHER 77
  - *Mid-Flood Weather* 79
  - *Other Models Fail* 84
- GROUP DISCUSSION QUESTIONS: 89
- ANSWERS: 89

.... 91

CHAPTER 3 | THE FLOOD 91

- PRE-FLOOD CIVILIZATIONS 91
- THE NARRATIVE 91
- INCEST 94
- BACKSTORY 95
- THE STORY OF NOAH 98
- ARK CAPACITY 100
- THE GLOBAL FLOOD 102
- GRASSLANDS 119
- POST-FLOOD WEATHER 121
- ICE AGE? 122
- TOWER OF BABEL 125
- ETHNICITIES 127
- CAVEMEN 132

Contents

GREAT FLOOD SYNOPSIS 135
GROUP DISCUSSION QUESTIONS: 137
ANSWERS: 137

CHAPTER 4 | "EARTH LOOKS OLD" 139

A POETIC INTERLUDE (AND ODE TO THE LORD) 141
CONFLICTING COSMOLOGIES 144
WHAT'S WRONG WITH "DEEP TIME"? 148
DISTURBANCE 149
POLYSTRATE FOSSILS 151
EROSION 157
FOSSILS AND FOSSIL FUELS 159
MUD FLOWS INTO ROCK 162
GREAT UNCONFORMITY 165
STARLIGHT 169
    Objection 1 .... *169*
    Objection 2 .... *170*
BONUS SUPPORT FROM BONES 171
GROUP DISCUSSION QUESTIONS: 173
ANSWERS: 173

CHAPTER 5 | RADIOMETRIC DATING 175

CARBON DATING 181
THE CONTRAST 184

CHAPTER 6 | SOME SCIENTIFIC DATING METHODS 189

THE SUN'S SIZE 190
COMETS 191
EROSION (AGAIN) 193
POPULATION CEILING AND DNA 196
THE LOW CEILING OF POPULATION 197
THE Y CHROMOSOME 199
GROUP DISCUSSION QUESTIONS: 204

iii

Contents

ANSWERS: 204

CHAPTER 7 | BIOLOGY 205

EVOLUTION 205
EVIL-YOU-SHUN 212
GROUP DISCUSSION QUESTIONS: 216
ANSWERS: 216

CHAPTER 8 | THE PROMISE 217

ON JESUS AND DESTRUCTION 219
RELIGION 221
RELATIONSHIP 227
WHO IS GOD IN THE BIBLE? 229
ISAIAH 53 231
WHEN GOD COMES BACK 238
HUMANITY 2.0 239
THE MILLENNIUM 240
JUDGMENT 243
CONCLUSION 254

CHAPTER 9 | MODEL SUMMARY & Q&A 259

CORE PREMISE 259
   1. Pre-Flood Earth *259*
   2. The Pre-Flood Water Layer *260*
   3. The Impact Event: Trigger for the Flood *261*
   4. The Global Crust Collapse & Flood Mechanics *261*
   5. The Immediate Post-Flood World *262*
   6. The Ice Age: A Direct Result of the Flood *262*
   7. Post-Flood Human and Animal Migration *263*
   8. Radiometric Dating: The Great Reset *263*
   9. Marine Survival Hypothesis *264*
   10. Magnetosphere Disturbance *264*
   11. Cosmic Debris Hypothesis *264*

Contents

12. Evidence Emerging Today 265
Final Thoughts 265

QUESTIONS & ANSWERS 266

WHERE DOES RADIATION COME FROM? 266
DOESN'T RADIOMETRIC DATING DISPROVE THE BIBLE'S TIMELINE? 266
WHERE WAS THE WATER BEFORE THE FLOOD? 267
WHY ARE THERE FOSSILS? 268
WHAT WAS EARTH LIKE BEFORE THE RUIN? 268
WHY ARE THERE ETHNICITIES? 269
WHY ARE THERE DIFFERENT LANGUAGES? 269
WHY ARE THERE SO MANY RELIGIONS? AND WHY ARE THEY SO SIMILAR?
.... 269
ON RELIGION 272
WHY DO PEOPLE AVOID CHRISTIANITY AND JUDAISM, IF BOTH ARE SUPPOSED TO BE GENUINE? AND WHY DON'T CHRISTIANS AND JEWS AGREE?
.... 274
WHY DOES EARTH LOOK OLD, ERODED, AND BROKEN? 275
HOW COULD A 40-DAY RAINFALL FLOOD THE WORLD, COVERING THE HIGHEST MOUNTAINS? 275
HOW COULD ONE MAN PUT ENOUGH ANIMALS ON A BARGE/BOAT/BOX TO GIVE US THE SPECIES WE HAVE TODAY? 277
WHAT HAPPENED TO ALL OF THE DINOSAURS? 278
   Leviathan 278
   Behemoth 282
WHY IS THERE STILL ICE ON THE POLES IF IT'S ALL MELTING? 284
HOW COULD A CRUST LAYER BE SUSPENDED ON A WATER LAYER? 285
WOULDN'T THE WATER LAYER BOIL AND CAUSE HEAT BUILDUP AND PRESSURE? 286
DON'T WE HAVE RECORDS THAT GO BACK MORE THAN 6,000 YEARS? .. 286
IS THIS ALL IN THE BIBLE? 286
WHY DIDN'T WE ALREADY KNOW THIS? 287
WHY DID GOD GIVE IT TO GARY WENTZ? 287

Contents

WHAT QUALIFIES YOU TO SPEAK FOR GOD? ARE YOU A PROPHET?...... 288
WHY DO WE NEED THE BIBLE WHEN WE HAVE INTELLIGENT DESIGN?.. 288
HASN'T SCIENCE PROVEN EVOLUTION AND DEEP TIME TO BE TRUE, THEREBY DISPROVING THE BIBLE?... 290

Preamble

What This Is

If you're really sensitive to hearing or reading about things of God, Christianity, and the Bible, and don't want to hear it, just put the book down now (or click off) and walk away from this. I promise to talk about God and His Word as I discuss His world—His broken and doomed world. ¯_(ツ)_/¯

I'm going to share with you what I think God shared with me in February of '23 (*this millennium*). He gave me an image in my mind of Earth (*see cutaways above and below*). It lasted just a few seconds. I had to mentally process what I had clearly seen for a bit in order to understand it better (*although the gist of it was quite clear to me—as were the ramifications*).

The Vision Itself

It was the 12th of February, and I was in my garage. I was actually contemplating the Earth and the Flood at the moment when a holographic-looking image appeared on my workbench in front of me. It was transparent but visible—a globe of Earth in a cutaway, about 14 inches tall.

As soon as I realized I was looking at a vision of the Earth, the image left the workbench and was strictly inside my mind but zoomed in (*it might have been all I could see, actually*). I could quite clearly see a deep layer of water on top of the brownish rock surface of the **Mantle**. The **Mantle**'s surface was smooth and brown (*intact*); the **Mantle**'s interior was not shown to me.

The **Water Layer** was encased in an outer rock ball, which was covered with a **Soil Layer** and some bodies of water. As soon as I understood what I was looking at, the vision faded away. As it disappeared, I felt a familiar touch on my head. It was my God letting me know that it was Him giving me the image (*I know His voice and touch*). The whole experience lasted about 5 to 8 seconds in all.

This message is a presentation of what came out of that brief vision and subsequent instruction from Him.

Preamble

Understanding the Vision

After receiving this amazing vision from God, I had to take a crash course on the fundamentals of geology and geophysics just to know enough to communicate this knowledge and its impact to others. Since reaching adulthood, my education on the topic has largely been self-guided anyway (*since I never attended a university*), but since the image came to me, God has taken me on a short study of how He made the planet for us.

This is the second impactful vision given to me by God. The first one was more spectacular and longer in length but more personal. I've seen other, less impactful visions before—like seeing Jesus while worshipping or thoughts that come to mind as images—but so far, I've only had the two really impressive ones.

This one has the potential to completely change man's understanding of cosmology, geophysics, and Flood geology (*it turns it on its ear, actually*). It demonstrates, very clearly and eloquently, the fact of **catastrophism** as opposed to **uniformitarianism**. (*Catastrophism says that the world was reshaped by one or more catastrophes in the past, while uniformitarianism says that the world has always been uniform, or constantly like it is now, with slow changes, if any at all.*) **Uniformitarianism is what is taught in schools, while catastrophism has been vehemently opposed by them.** Uniformitarianism is an attempt to remove God from the picture, while creationism keeps Him central to the message.

Young Earth Creationism & Science

When God feeds me with information or knowledge (*not necessarily with a vision*), it's usually little spoon-sized bits at a time. I have to show interest in whatever it is, or else He may not give me any more. This time, I paid attention, and He taught me starting from where my knowledge base was (*which is a lot higher now—a year into this*).

I have always had an interest in scientific knowledge but never attended university due to their anti-God dogma and the expense associated with enrollment (*not to mention making a six-digit income without it—and the debt*). My understanding, then, has been born and nurtured by God. This doesn't make me a genius, and I am sometimes quick to forget what He has taught me (*even less apt to apply it*). And He has only shared a little bit regarding a few things anyway throughout my adult life (*or maybe that's how much I have paid attention*).

And although I say that God gave this to me, don't think it means that I am some "holier-than-thou" kind of guy, or in any way superior—I most definitely am not (*just ask my wife*)!

Ever since the mid-'80s, I have depended heavily on the tutelage of **Young Earth Creation (YEC)** teachers and scientists for my knowledge in areas of science and cosmology (*the study of the beginning of the cosmos*). And YEC scientists

Preamble

are quite knowledgeable about the science of it all—at least as much as anyone else (*they're scientists, after all*).

In fact, the creationist view is gaining popularity in the world as people wake up to what I am about to discuss: the truth that actually fits with the evidence. It is the creationists who are making exciting discoveries these days in science, not the evolutionists.

And this revelation, though not a discovery, is very exciting to me. It opens up the whole history of the world to me and brings everything into its proper place—and timing. It is a **backstory** that has not been shared before, to my knowledge.

Why This Matters

Anyway, now that I understand better what the Holy Spirit showed me that day about the Earth, physically, I'm passing it on to you. That's what this book is going to discuss. And if I do this right, it will be very easy for you to understand my message and determine whether or not to wrap this new knowledge into your own understanding. To be clear, this is a message from me to you of a special vision from God to me. Sound fun?

How To Take This

Unfortunately, many people today don't really like discussing certain subjects; like Religion and Politics. Well, at least we can dodge one of those topics here. Politics is far too petty a subject for this revelation anyway, and I can only take so much stupidity at a time. (*Sorry, but politics is filled with dishonest people acting stupidly and selfishly. Well, religion is too, but this needs to be discussed; it's far too important to ignore.*) And know that the enemy cares deeply about this topic. It's been one of his priorities from square one. That's probably because the stakes are so high for all eternity; not just this voting cycle. So, we've just got to discuss this.

You know, the enemy has never quit using the flood against the Lord; from as soon as it happened. Then, when everyone was split up from Babel a hundred years later by God, the demons went to work right away; using the pain of it all to force a wedge between Him and us. They became the new gods to many people worldwide.

We're going to go BIG, and will cover no less than:

- How we got here;
- The state of the Earth now and why;
- What its condition will likely be going forward;
- And not just how its state affects humanity, but how humanity affected the Earth and all other life upon it as well (*past, present, and future*).

Would you like to go there with me? It shouldn't hurt too much, unless you allow your pride to get hurt. Learning new points of view can be lots of fun, with the right way of approaching them. I don't claim any authority over anyone, so everybody is free to take this or leave it as they see fit. But this topic is controversial. It's actually very controversial to some. In fact, some folks are really not going to like what I have to show them here. I can try to be sensitive to their hot buttons, but someone is going to be instantly upset with this. I hope that isn't you. Luckily for them, I have no credentials, so they will find it very easy to pooh-pooh my model and go on about their merry, uninformed way.

I am a Young Earther and have been my whole life—before I knew the term for it, even. This point of view that I am presenting fits squarely within that framework, since I have never had any other point of view. Oh, I've listened, and read, and taken the tests regarding what I call "Olde Earthe" stuff; I've just never been convinced by the arguments of Deep Time (*m/billions o' years*) or Evolution (*evil-you-shun*). Not for a minute. And the title and philosophy of "Young Earth Creationism" is in response to the teaching of evolution (*from amoeba to man, said to occur over millions or billions of years*) as it has been and is being forced upon us from the media, entertainment, education, the science industries, as well as the government (*any government at any level, worldwide*). All power bases involved in propaganda or information release seem to be spreading evolution occurring over m/billions of years as a forgone fact. But is it? Is this world really billions of years old? Can you prove it? Can anyone, regardless of position, prove their case?

Why Me?

Thanks for joining me on this journey to the center of the Earth and back again; then on past us, right into eternity. We're going to look at the best-selling book of all time, one of the oldest, the Bible, and see what it says about the age of the Earth and how it was put together by the Master Designer and Builder. We're also going to unpack that short vision and lay it out with all of its consequences, both scientifically and theologically.

But before we jump right into the details of this amazing vision and its impact, I want to take just another minute and consider the context, or perspective, from which I come into this. I am, after all, anti-religion (*which I consider a way of telling God to pay up*) and I consider the Bible to be true; not a textbook for world dominion (*as some have attempted to make it*).

I'm perfectly sane; I just don't believe the nonsense that the overwhelming majority of people in the world have been taught about This Broken Planet (*with most of them falling for it*). It is one of the biggest hoaxes known to mankind and I have simply refused to buy into it. And if you will be honest with yourself, you should realize that you have some doubts or questions about it too that still linger. (*Unless you just know what it is and keep using it as a weapon against the Bible thumpers and others who hold to a strict religious worldview that you hate.*) If that isn't you, then stick with me and I'll show you a whole new way to look at planet Earth; This Broken Planet.

Before I share anything on why God may have picked me as someone to give this to, I want to point out that it is not because of my accomplishments, character, or lineage (*to my knowledge*). There is nothing about me that is outstanding as a person. I am not particularly inspirational to others (*unless by sharing God's message alone*). I certainly cannot claim being better than any other person that I know (*as in being a good, righteous, or holy person*). And I have no idea how many other people He gave this revelation to either (*I only know of one othe*r). Maybe I'm the millionth; maybe I'm only one of two. Anyway, here's how I see God's heart concerning who He likes to use. Really, the message is about the message; not the messenger. Quickly, here is an example of a message and how it is received and presented. See if you can make the connection.

> 1st Corinthians 1:18-29, (NLT)
>
> The message of the cross is foolish to those who are headed for destruction! But we who are being saved know it is the very power of God. As the Scriptures say,
>
> "I will destroy the wisdom of the wise and discard the intelligence of the intelligent." (Isaiah 29:14, NLT)
>
> So where does this leave the philosophers, the scholars, and the world's brilliant debaters? God has made the wisdom of this world look foolish. Since God in his wisdom saw to it that the world would never know him through human wisdom, he has used our foolish preaching to save those who believe. It is foolish to the Jews, who ask for signs from heaven. And it is foolish to the Greeks, who seek human wisdom. So when we preach that Christ was crucified, the Jews are offended and the Gentiles say it's all nonsense.

> But to those called by God to salvation, both Jews and Greeks, Christ is the power of God and the wisdom of God. This foolish plan of God is wiser than the wisest of human plans, and God's weakness is stronger than the greatest of human strength.
>
> Remember, dear brothers and sisters, that few of you were wise in the world's eyes or powerful or wealthy when God called you. Instead, **God chose things the world considers foolish in order to shame those who think they are wise.** And he chose things that are powerless to shame those who are powerful. God chose things despised by the world, things counted as nothing at all, and used them to bring to nothing what the world considers important. As a result, no one can ever boast in the presence of God. ~ Paul, Apostle of Christ

So, why me? I don't really know; does it mean that I'm just pathetic enough for the job? I'm a nobody; so the message isn't about the messenger, that's for sure.

In God's Image

I feel it would be helpful to share my own personal perspective and understanding concerning the Creator as we go through this. So here's a quick bit about us being designed and made in God's image.

I think that I can add one more way to the list of ways in which we are made in the image of the Living God: He is infinite, which means that somehow we are inside of Him. (*I'll*

Preamble

actually expand on this in chapter one.) We are not infinite; but we do have other beings inside of us also.

We've learned through scientific research that our bodies have amazing autonomous structures in them that seem to be mechanical in nature, while also being chemical and biological, I guess. But what struck me about some of these is how they seem like tiny creatures setting about their appointed tasks (*like white blood cells, for example, searching, sniffing and snuffing out infections—or the microbes that are the intruders*). In fact, we are all made up of tiny little particles that form into things that have function and purpose. Consider again the immune system in your body. It is somewhat autonomous, and yet it works on behalf of, for the betterment and benefit of, your body.

Or just the cells, themselves; I've seen video of two heart cells filmed on a dish in a lab that start beating together when they come into contact each other.

Or in your gut, your intestines, you have many kinds of cultures or colonies of tiny microorganisms. Some are beneficial and necessary, while others are the intruders, which the guardian cells of your body are fighting off. (*Eat those veggies, kids.*)

It seems that everything about your body, from that furry skin of yours down to the innermost, infinitesimal structures of your cells, says that you are just one big machine, made up of the tiniest things in creation. And then when they come together to form your brain…wow, I'd imagine that's the most complicated thing in the universe! It's also one of the most fragile, in my opinion.

Not a big point, I know, but I just wanted to mention it. The human body is indeed a miraculous work of pure, infinite genius. And that's just the physical part of us (*the body*). There are also the other, ethereal parts that are even more amazing. We are all made up of three parts: body, soul, and spirit. Maybe I'll dig into this later.

Fragile Hearts And Minds

When we were little, we had all kinds of things shoved into our little, fragile, malleable, tender, and oh so impressionable minds. I was one of the lucky few who grew up with God's Holy Spirit in my brain and heart (*so to speak*) from my earliest knowledge or awakening. And I never really pushed Him out, even in my most disobedient years. So He has been with me the whole time; leading and protecting me, even when I disobeyed (*numerous, nay, countless times*). All I have to do is come back to Him and show Him repentance; then He always restores (*His mercy is new every morning*). I have always tried to be more loving and loyal to Him above all others. He is the only God I have ever had.

One way we can protect our hearts and minds from deception is to verify the message we are receiving with what we already know about life from the Word of God—the Bible. This is in opposition to the message of the enemy, which says to stay away from it; they say "get it out of the schools, and away from the children!" "Separation of Church and State!"

Preamble

And if the enemy doesn't want us to know it, then we must need to know it. If he's keeping it from us, then we must need it in some measure someway. And in the case of the Bible, we need it a lot.

So I invite everyone to put this Flood theory up against what we can actually see in the world beneath our feet (*through science*) and to put it up against the light of the Bible (*through theology*). Do that. With everything. Always. You won't go wrong.

Again I say, don't get involved in foolish, ignorant arguments that only start fights. A servant of the Lord must not quarrel but must be kind to everyone, be able to teach, and be patient with difficult people. Gently instruct those who oppose the truth. Perhaps God will change those people's hearts, and they will learn the truth. Then they will come to their senses and escape from the devil's trap. **For they have been held captive by him to do whatever he wants**. (2 Timothy 2:23-26, NLT)

Jesus helps people escape the clutches of Satan (*enemy of both God and man*). And when you have been set free by Jesus, you are freed indeed. Religion, "deep time" (*m/billions of years*), and "evil-you-shun" (*my term for the evolution theory*) are traps set by your enemy. Although you can be freed from his power, you are never free of the enemy's influence for as long as you wear that skin suit you're in. The test is always on. The clock never stops until it's over. His snares are easy to fall into when your eyes are off of your

Creator *(for just an instant)*. I think that is why Satan is still alive and able to interact with us today; to see if he can pull us away from Jesus and the Father and the Holy Spirit (*and their Bible*); to get us to deny Jesus before men. Or at least to be ineffective or dormant in God's work. Oh, and getting us to hurt each other seems to be a favorite trick of the enemy's too.

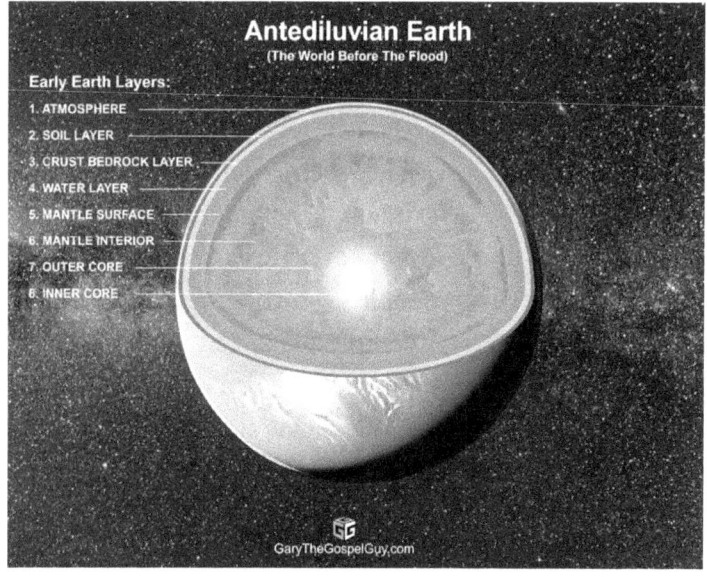

Find the Water Layer in this cutaway picture of Earth and see where it is, relative to the other parts. It was located between the Mantle below it and the Crust Bedrock above it, before Noah's Flood of the Bible. When it broke, the world fell into the scalding hot water, quickly turning to mud, maybe miles thick. If you picture a melon cut open, it helps to imagine the Mantle of Earth. Above that is a lot of hot water (heated by the Mantle) and that is encased in a stone ball Crust.

Chapter 1 | Competing Worldviews

Cosmology

When you ask the question, "How did we get here?" you're engaging in cosmology—the study of the beginning of the cosmos. The cosmos is the physical universe—all physical matter. (*For me, cosmology is also the study of the beginning of spiritual things and spirit beings as well.*) There are good cosmologies and there are bad cosmologies. But the word "cosmology" is neutral. It's just the arena for the fight, so to speak.

It's also a belief. What you believe about cosmology is a matter of faith; not proof—regardless of your position. There is no scientific data from the creation and years immediately following, enabling us to say that science can prove a cosmology is false (*see chapter 6*), it can only speculate—possibly really well, if done right.

If you believe that science has all of the answers, including the opinions stated by scientists, then you are exercising your faith in their comments. Your faith is in ***them***. How is that superior to believing the story put forth by a Book that claims to be from the One Who put the Earth in place? A Book that is true in all of its teachings, which are many and varied, just as life is. That's a Book worth investigating, in my view. As are the claims of scientists—it's all worth investigating, isn't it?

(Did you answer my question as to which is superior by looking at the evidence? Did you answer that the secular view is superior because it comes from science?) Well, let's do that together. Let's see whose claims make the most sense to us scientifically, logically, and theologically (*as we put our fears, biases, hatred, and any other negative feelings aside*). "Father, help us to open our hearts and minds to your truth, amen."

The Bible is not Christianity; it's a book that says it's from the Creator of all creation. Christianity has become a religion. Christianity needs to be monitored for truth. The Canonical Bible has already been vetted and is trustworthy (*just watch out for a few bad translations out there. See chapter 8 for a short list*). People—*what the Church is made of*—are fickle; one minute they're fine, the next, they're not good at all. Don't judge The BOOK by its COVER (*that is, the ones smothering it; those calling themselves the Church*). It teaches us to have self-control; not to exercise control over others, for example. Whoever has done that in the past or today, was and is wrong for doing so. Much abuse has come through religious Christianity over the centuries since becoming mainstream. And there is more to come; no doubt.

How you answer the question, "How did we get here?" will weigh heavily on how you receive this new theory of mine. And it is just a theory, since I am not God, or even a

1 | Competing Worldviews

prophet of God. I'm just a regular guy with a high school diploma. But the bottom of it will reveal your willingness to accept or deny a spiritual reality—regardless of the ample amount of scientific data in the world that accompanies it—*there is a God*...and He talks to us.

You see, this kind of thing will always get boiled down to faith, regardless of the evidence before us. We all have the same evidence, you know, what differs is how we receive it, or allow ourselves to perceive it. This book is my public profession of my own faith (*my cosmology*). I think it's worth sharing.

If you didn't know, the word "cosmology" is a word used in both education and religion. And they each seem to have their own opposing in-house cosmological views nowadays as well. Like I said; it's controversial.

Sadly, we've been born into a war zone, and the nannies that have been watching over the orphans have been prepping them for slaughter. The schools have our children and grandchildren so confused that they can't even define what a woman is, let alone treat her right. I see where the schools have been taking our children. They are the proverbial "Pied Piper", leading the children to death everlasting—as are the religions.

At one time, the local college was teaching what the local church said about scientific subjects. But now, they differ from each other, and there are differences of opinion inside each institution's walls as well. Neither is the hero; both have been villainous. (*If you are a church attendee, then you know that Jesus is the hero, not those in the pews or pulpits.*) And I'm not saying that teachers or pastors are villains. I'm saying that the ideologies coming from either can have *error*, which is the real villain.

Just as scientists disagree on much of cosmology with each other, the Church members disagree with each other about it too. But somehow along the way, much of the Church decided to follow the college guys instead of God. They gave up on God too soon, before all of the evidence was in. No wonder disagreements abound.

The Past: Abstract or Reality?

Well, the past is reality that fades into abstraction in our minds. We can't really keep ahold of time as it slips past us. But God can. He has a record of every thought, word, or deed of every person ever conceived written in the books of life.

We tend to think the way that we are taught to think. Luckily, one of my closest life teachers growing up was a peaceful, humble scientist and genius—my dad. My other most prominent teacher was my mom; whose faith is rock solid and she has a good sense of reality as well. They're both very grounded. They both taught me **how** to think, not necessarily **what** to think. They both showed me to God, like all good parents should.

If you were taught primarily by schoolteachers, you may have picked up some bad information. When you are under a

1 | Competing Worldviews

teacher, you are at the mercy of their biases and worldview (*and those of their superiors if in an organization, like school*). There is much capacity for truth to be abused in a formalized setting, such as school. Well, no worse than any other setting, it is just usually mass communicated from a school, impacting more souls at a time (*like broadcasting fake news*). Well, there is the demand of regurgitation of the so-called "facts" that they present and grade you on, too. BTW, the same can be said of churches when it comes to spreading fake news and views. I have yet to find an organization made up of people that has true purity in its message.

Now I know that schools in the world are run by someone with an agenda against our kids. That someone uses people without their knowledge of even being used! He is a very smooth operator. A lot of teachers have had to answer to God for telling lies to children that are harmful to their relationship with Him, even though they were deceived as well.

The real deceiver is Satan, but people are his instruments of delivery. And God will make us answer for listening to and following the devil instead of Him. God has been pretty upfront about that the whole time. (*He still hasn't forgiven the snake for letting Satan in, I guess.*)

But beyond that, we tend to put the past into a more abstract context in our minds. Even our own memories can be deceiving or unreliable. As a collective, we have had a tendency to stretch out our recollection of history, making it seem longer than it really was. I mean, it's so abstract that we can apparently stretch it into millions of years in our conceptualization of it, and beyond that even into billions, which if discussing years is just crazy. God gave us logic to deduce that such a concept is ludicrous if applied to the world beneath our feet today. But hey, if you're going to Nah Nah Land, you might as well go all the way, I guess. It really does cost one's grip on reality to cradle the notion of a million years of history for us here, though. Either hang onto reality or millions of years—you can't hold on to both. They're too far apart. (*See? I am a diehard Young Earther. Not to worry, that's about as abusive as I get.*)

The reality is that our history on this proving ground has been very brief and it is due to be fulfilled in a very short time (*1,000 more years to go*). Satan's rule over us (*with his lies*) is about to end. That is because Jesus is about to come back and imprison Satan and his entire army. He will destroy the kingdom of darkness and rule God's Kingdom of Light from Jerusalem. That's when heaven comes to Earth.

My Cosmology

What I believe about the world has always started and stopped with the Bible. And no, that is not circular reasoning; it's where I start and where I finish. It's the first book to come off the shelf and the last to go back on (*so to speak, I really don't shelve the Bible that I use daily*).

The reason this practice of mine has endured my entire adult life is that the Bible continues to survive all assaults against it. If it could be ruled out, I'd let it be ruled out. But

1 | Competing Worldviews

that just hasn't happened yet with respect to Earth's origin and the Bible's validity in all matters.

In fact, when I hear the theories of how Earth came into existence that are contrary to scripture, they just seem silly to me (*whether religion- or science-born*). But I have always listened to and pondered them open-mindedly. I respect those people, as individuals, regardless of their views, just for being made in the image of the Living God. Although the opposite is true regarding my Y.E.C. beliefs and others who hold to the same teaching as the universities—they laugh in my face. But it's okay; I understand why they act as they do. And the same can be said of those who are less than cordial when presenting a biblical point of view to others who do not. We all need to work harder at being nice to each other when disagreeing. I am speaking of person to person here, not institution to institution, or kingdom against kingdom, or ideology versus ideology.

All I am trying to say is that I am not judgmental toward those who see things differently than I do. And I hope that the reader (*that's you*) can be open-minded toward my message here as well. But it might get a little weird for you, depending on what you already believe about this stuff. I'm pretty sure that you've never received this message before now—either in church or in school. It doesn't matter how smart you are, how well-read or not, or any other distinction that could be made about who you are and what you know. No matter what, you are on level ground with everyone else in this because nobody else (*with few exceptions*) has received this message either (*from me*). This is the world premiere of this cosmological view (*actually, the online version was published first*).

If I could offer any spoiler at all, it would be that the lies told worldwide for over these last 200 years have been hugely, and I do mean **hugely**, exaggerated from reality. So that would make my idea seem like fantasy to those who have heard and believed the biggest fantasies of them all that are godless.

An important thing to remember as you read through this presentation is that you will be presented with facts about the world—the real world. As you receive these facts about Earth, and see how they fit with what the Bible says and not with what the schools say, try to realize that the rest of the Bible is true to observable science as well (*when applicable*).

The Bible can be trusted in every single message it puts forth. It is through studying the world that the Bible comes to life; and it is through studying the Bible that the world discovers life.

The Spiritual Side of Things

This, too, may be strange for you, but I'm about to explain a very personal, but simple, scenario that I have in my mind regarding the creation and how God did it. But this also touches on the very essence of God and Who He is. It also discusses Jesus and the Holy Spirit in a way that might be new to you. But trust me, I'm not leaving the Bible in any of this. At least, I don't see me as getting unbiblical here. Before reacting to some of these details, please consider their

1 | Competing Worldviews

agreement with biblical revelation. See if what I am saying is biblical. Test it.

God

God is infinite and lives in infinite light. But why don't we see His light? If it's infinite, we should see it. But we have darkness in space. And the light that we have is not like the light that He calls His glory (*a special, spiritual light that is not broken by objects the way our physical light is—it just shines right through anything physical*).

His light casts no shadows and is actually unapproachable.

> [God] alone can never die, and He lives in light so brilliant that no human can approach Him. No human [mortal] eye has ever seen Him, nor ever will. All honor and power to Him forever! Amen. (1 Timothy 6:16, NLT)

If we were to try to enter God's presence, we would cease to exist, I guess, but how would we even get to Him? He is literally on His own. No mortal human may enter His unapproachable light. I call this His **Divine Realm**; a place where He and only He exists, since He is infinite and His creation is not.

I believe it is the evil that we carry in us that cannot exist in the presence of God and His Unapproachable Light. Evil is in us like a virus, but a part of us, like our physical DNA. I call it our spiritual DNA (the ***Depraved Nature of Adam***), since it's as incurable as ridding ourselves of our deoxyribonucleic acid (*the physical DNA that cannot be removed from us if we want to have a physical body*). So what is God's solution to this apparent problem (*of creating creatures that He knows will quickly become imperfect and therefore unable to approach Him*)?

    Well, God had to place a kind of buffer between His unapproachable goodness that would automatically destroy our evil and us along with it. I mean, it probably didn't take Satan any time at all to fall from God's grace (*thinking he could be equal to God*). So this buffer had to be in place ahead of time—before the creation. And since God knows all things from start to finish and then some, He knew that this would happen, oh, and that Adam and Eve would follow suit, just as quickly. God created beings that were good but they became bad on their own. We ruined ourselves. He is not to blame for our ruin, especially since He has been protecting us all along.

    So, He made a buffer between His creation and Himself. This buffer is a living Being—a part of Him that is like His arms and hands (*as a metaphor*). God divided or multiplied Himself without losing any of His presence in each Person that He presents to us. He actually gives Himself two new Personas or Persons within Himself: He calls One His "Son"; and the Other, His "Spirit"—The Right Hand and the Left Hand of God, if you will. He is filling Both and giving Each His own status as God over creation (*that is soon to come*). They are All considered God—One God in three Persons.

1 | Competing Worldviews

Together, I call Jesus and Holy Spirit "Adonai" ("*my Lord*", *in Hebrew*). For me, They are One and are unified. For me, there is but one Adonai. Jesus is His human Body and Soul; Holy Spirit is His Spirit, Who can come into contact with me without destroying me. At the same time, Adonai is very much connected to the Father and shares His position above angels and us. They are together very much our God; the One Living God. Using logic, three infinite Persons must be united, by default. How could God not be One with and within His infinite Self ($x3$)? (*1 x 1 x 1 = 1. For producing 1 God; not 3. He didn't add to Himself; He multiplied or divided Himself*).

I am made in His image, partly because He is One Who holds many lifeforms inside of Him and I am one who holds many lifeforms inside of me (*physically*). *You too.*

A Spiritual Image of the Creation

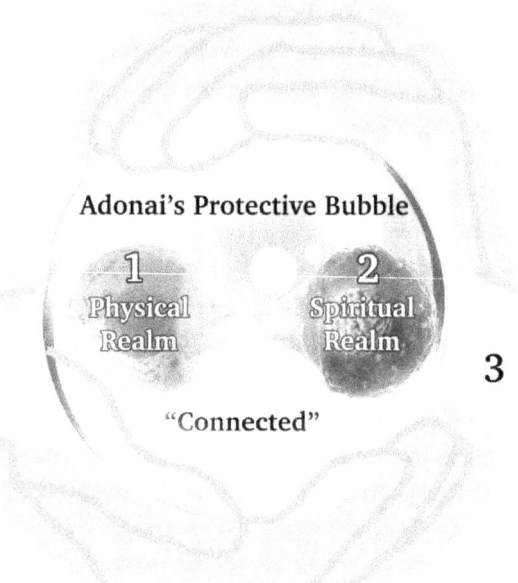

1. Physical Realm 2. Spiritual Realm 3. Divine Realm
The hands are Jesus and Holy Spirit; 3 is Father God

I have in my mind this God with His Two Hands cupped together before Him. Like in the image of the hands. Inside the cupped hands, He begins to create the two realms that are the creation: the Physical and the Spiritual Realms. The cupped hands are to protect us from His righteousness.

1 | Competing Worldviews

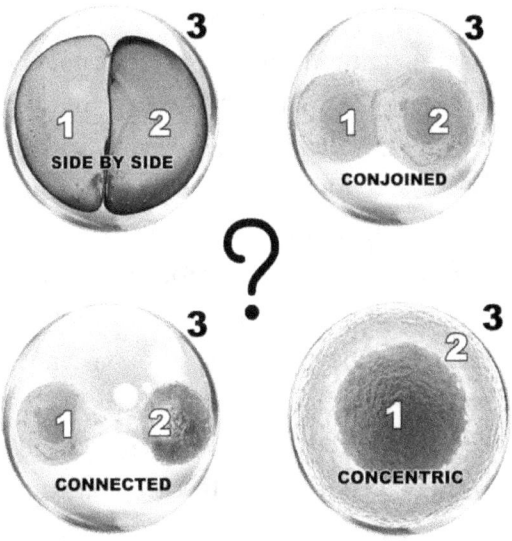

Other possible arrangements or configurations of physical and spiritual realms

Possible Arrangements of both Realms: Side by Side; Conjoined; Connected; Concentric; Something else (*phased*)?

Inside these hands, that are larger than all creation, He begins to form matter and spirit. I see streams of both flowing from His Hands into the creation, forming and shaping it all in a fast, fluid manner. Like a beautiful sculpture that forms itself. The Spirit was hovering over the waters. Before it was even all together, He began breathing life into it; abundant life; numerous and varied. And last of all, He made us in His image. When you think of it, woman was the pinnacle of His creative works; the icing on the cake. *Try to stay humble, girls.*

So really, matter didn't come from nothing; it came from Him. And we are inside of Him at this moment. But we have

Jesus and the Holy Spirit between us and Him (*Father God*). They are our protection from the Father's righteousness that would annihilate us. He put them there for our benefit; not His. If we did not have Adonai between us and God, we would have been destroyed immediately, I'm sure. And since we must go through Adonai to get to the Father, and because Father says so, we call Jesus and Holy Spirit "God". And from where we are standing, they are God; our Creator. You may argue that the Father brought them into existence and thereby created them; however, He then proceeded to create us through them, making them our God. So their divine relationship is above our station to even contemplate. But He didn't really bring them into existence from nothing, they are and always have been a part of Him. And we are to worship both Holy Spirit and Jesus as God. It's a command. They are shown to be God and we are to worship God. That's it.

And think about this: All that Jesus and the Holy Spirit and the Father do is for us. God is totally focused on us—each and every one of us personally. He has done nothing but serve us the whole time.

(*If this comment triggers some pain in your heart toward the loss of a loved one or some other tragedy that brings animus toward God from you, please remember that He is not the one who brought pain into the world; we are. And when you turn from Him in anger, He is unable to soothe your pain. Better to turn toward Him in your anger. Some things just need to be confronted and dealt with. He's actually easy to talk to about such difficult matters. And He really does understand*

1 | Competing Worldviews

and love you through it all. Please give Him that chance. And remember that if we are His, the pain we have now is nothing compared to the joy that awaits us with Him when this world has ended.)

> For in him [Jesus] all things were created, things in heaven and on earth, visible and invisible, whether thrones or dominions or rulers or authorities. All things were created through Him and for Him. He is before all things, and **in Him all things hold together.** (Colossians 1:16-17, Berean Standard Bible)

This describes the physics that we cannot yet figure out. It says that all things are held together by God (*namely, Jesus*). Think of the force that keeps an atomic nucleus intact.

String theory seems to be searching for a mathematical model to explain God's power. There might be one out there, someplace. Maybe God will give it to someone if they ask for it nicely, instead of making up stuff like He isn't even here in the mix.

Personally, I don't agree with these theories of physics:

- String Theory
- Dimensions below or beyond our 4 (*time is the fourth*)
- A big bang

We don't really need to explain the unidentified source of energy that keeps atoms intact (*beyond neutrons*), for example (*but it would be nice*). However, we do know that they're held together by God—one atom at a time—because He says so and He can do that. If you want to try to figure it out with math, go ahead and good luck. Just don't expect me to go along with something that has nothing to do with God, especially if it tries to remove His influence. And quit trying to leave the time continuum. Wherever you go, you're in time (*even in heaven or hell*). You can't escape it. Going back and forth along the timeline would be neat, but if possible that would be probably more of a spiritual exercise than a physical one (*guessing*). However, God can take us wherever or whenever He wants to (*not guessing*). "Will He?" is another question.

My explanation of life is much simpler than the cheesy theory of string, anyway. As already shown and discussed above, there are three realms of existence:

1. Our Physical Realm (*the universe*);
2. the Spirit Realm (*where angels are from*);
3. and the Divine Realm (*which includes God Himself*).

Time covers both created realms and spirit can control matter (*at least, my brain thinks so*). God does not have a habitat; such a thing is for His creatures, not Him. He *is* His habitat (*ours too, in a way*). Of course heaven is inside of time. Time is a part of the creation. Heaven is a part of the creation. Time will soon run out in heaven (*just about another thousand years to go for that place too—the whole Spirit Realm*). If it will end, then it had to have had a beginning as well.

1 | Competing Worldviews

It's interesting and fun to learn new things about the world and universe we are in. But as we explore new ways of understanding or seeing things, we need to remain grounded to the Word of God, since that is where truth lives. Talking of things that don't exist as though they do is called superstition. And that is never a good thing when deception is involved (*and Satan sees to it that all things are touched by deception*). But when God lets go of the atoms He's holding together, that will be a big boom! Well, He did say that He'll destroy the creation with fire. That will be a big fireball to behold, indeed.

I don't worry about physics because it's all in His Hands. And He is just as dependable as physics is (*well, infinitely more dependable actually*). I see physics' laws as being a testament to His reliability. Physics is reliable because God is reliable. He established those laws and keeps them secure. I can say that gravity keeps me grounded as easily as I can say that He keeps me grounded. And I could mean the exact same thing with each statement. My body is not an atomic bomb, only because He is still holding my body's atoms together. His supernatural power feels more natural than anything because it is the foundation of everything that is "real".

Biblical Cosmology

The Bible is definitely not a science textbook, but it has no problem treating the most complicated of topics with a simple approach that is very matter-of-fact and elegant. God's creation isn't described for us the way that a textbook would attempt; it just says that He created everything by speaking it into existence. And it's very different from the myths and legends of various religions and cultures in a literary sense, while having some connection to many early beliefs in some formative ways (*like having a good man being saved with a bunch of animals in a boat, or just the mention of a worldwide flood*). But there is no other book like it in terms of its message.

Genesis is written in the narrative style; not poetry. Ask any Hebrew scholar and they should back me up on that. And not just in a literary sense, but also a factual sense—its truth shines through. Through a bit of personal study into these scientific things, I have learned that the Bible doesn't waver from truth.

Very little detail is given about the actual composition of the Crust, Water, and Mantle Layers of Earth. In one passage it just says that He separated the water from the land. Peter said that the land came up out of the waters in the creation. This theory that I am proposing is not in the Bible, per se, it is something that God showed me directly—but fits with the Bible's account exactly. Here's how Peter put it:

> They [Scoffers] deliberately forget that God made the heavens by the word of his command, and he brought the earth out from the water and surrounded it with water. (2 Peter 3:5, NLT)

So how cryptic is that compared to giving a detailed explanation of what it looked like exactly (*with graphs and pictures and CGI video*)?

1 | Competing Worldviews

Really, the details aren't that important. That's what science is for—getting to the details of it all—and Peter was more concerned with putting up with scoffers and their scoffing at us for our beliefs (*and holding onto them through it all*), than the science of Earth's composition.

But we have to admit that this account by Peter does describe what I'm showing in this presentation (*I mean, you should when you see it*). If we hold in our minds an image of the Mantle covered by a Water Layer, that is encased in a rocky Crust Bedrock covered with a Soil Layer with freshwater lakes and seas, then this seems to be right on track. And the atmosphere is infused with water too (*especially then*). So my idea here matches exactly what Peter said; water is on both sides of the land (*above and below; surrounding it*).

> Then God said, "Let there be a space between the waters, to separate the waters of the heavens from the waters of the earth." And that is what happened. God made this space to separate the waters of the earth from the waters of the heavens. God called the space "sky." And evening passed and morning came, marking the second day. (Genesis 1:6-8, NLT)

I believe this to be a description of the creation of the atmosphere, separating it (*distinguishing it*) from the liquid water, and establishing the boundary between the two—surficial and atmospheric water; liquid and gas—we call it "sea level". This would have been the discontinuity between the two forms of water on the planet's surface.

Let me play with the text just a bit to show how we might phrase this today, knowing what we know about atoms. This is just to illustrate how wording can change over the years when new knowledge comes to us. See if this will help to see this verse in a different light.

Then God said, "Let there be more space between the water molecules, to distinguish the moisture in the air from the moisture in liquid form." And that is what happened. God made this atmosphere to distinguish the liquid water from the gaseous water. God called the air "sky".
(made up version for illustrative purposes)

Obviously, the words I'm using are full of 21st century understanding and concepts that the overwhelming majority of humanity would not have had any clue of. How could anyone before the discovery of atoms have known about this concept of stretching out a water molecule to make it gaseous, using the same atoms?

The bottom line to biblical cosmology is that God (*an outside force*) made everything that exists from absolutely nothing outside of Himself. This is very different from the atheistic rebuttal that everything just came into being without any outside force or catalyst at all. Of the two, which idea sounds more likely to you: Something out of nothing with nothing acting upon nothing; or something out of nothing with a Supreme Being doing the work, with an amazing intellect and power that cannot be fathomed? It's also different from the notion that God *is* His creation and His creation *is* Him—

1 | Competing Worldviews

as in being one, like in Pantheism or Panentheism. (*These two thoughts don't see the distinction between physical matter and spirit, I guess.*) God made matter but is not made up of matter (*except for the physical body of Jesus*).

It's also very different from the secular notion of a Big Bang, that doesn't really even address where the matter and energy came from that was supposed to have already existed.

Now, the Bible says that God spoke everything into existence and does not describe what I have described here that I see in my mind (*swirls inside of Hands*). What I described above is what I see in my mind for understanding; not what I read in the Bible. Are they different? Yes. Are they compatible? Why not?

To Whom was God speaking, anyway, during the creation account of Genesis one? Was it not to Jesus and Holy Spirit, through Whom He created all of creation? Are They not part of Him, yet distinct? Who else was there? Not Satan, or Michael or any other angels. Not Adam or Eve. Not creatures. Nothing was created yet. Jesus and Holy Spirit appear to be different forms of God Who do not destroy us immediately upon contact because of our inherent evil nature, while maintaining their connection and equality with God (*at least, for us they're equal and not fully separate*). They are our connection to God, the Father, and They protect us from His righteousness.

Precision

Another important factor in all of this is not only that the creation exists but also that it is extremely complex and ordered with design that shows up on all levels—even those we cannot see unaided. And not only that it is extremely complex but also that it works in a very narrow band of possibilities, as in, it's been finely-tuned exactly as it is—from the infinitesimal to the immense (*atomic to astronomic*). I should say "as it was before the Flood", since the Flood completely changed the world from its original state at the creation.

All things show irreducible complexity—meaning it cannot work without all the parts being in place together from the start. It's "all-at-once-creation" or nothing. And we can see it with our own eyes, too. Our minds understand this when the enemy's clouds are not in our heads.

The likelihood of such a complex design springing up out of nothingness is just too out there for me to take seriously. I actually NEED to have a God that is bigger than life to create life. Nothing else can make any sense to me at all. Sorry if you don't get that.

So in my biblical cosmology, a Creator created creation complete and finished, with nothing left to complete when it was finished. He did this in just 144 hours (*6 days*), literally.

Why did He take so long? (*He could have done it instantly.*) Why did He create things in an order that excludes

1 | Competing Worldviews

long gaps in the process of creation? (*To keep His way the only possible way. Instant creation with "irreducible complexity" is real, whether in the cells or the solar systems.*) Why didn't God use evolution to create life? (*Because it goes against His design. He doesn't leave things to chance. He meticulously designed each creature the way it is to show off His genius. Creation shows God's way; evolution is just an attempt to take God out of the way. It's a dumb idea.*) Why does the world look so old and beaten up? (*See chapter 4.*)

Hang in there and I hope to answer as many questions about it as I can.

Model Preamble

Wouldn't it be great if we could find an explanation that we could all get behind and investigate further? A cosmology that actually fits the observed conditions of Earth? That would be a fresh breeze of thought, wouldn't it? (*As opposed to the mainstream view that is farther off than you might imagine?*)

I hope to offer that to you in this presentation. If it doesn't fit with your scientific approach to our existence and Earth's state, then hopefully you will be able to kick in some critical thinking and big-time logic to see which view is more accurate with our observations worldwide. (*Understanding of physics helps too.*)

And then there's the challenge of dropping tightly held biases that go deep into the psyche. Give that a shot.

There's a good chance that I might challenge your spiritual view of our existence. (*If you don't have a spiritual view, then take a gander at mine and see what you think.*)

I hope that you will be able to have your faith in God and keep your sciences too (*as I do*). I have no conflicts at all between my faith in God and His Bible against my understanding of science (*actual science; not theories of scientists*). They are harmonious with each other. And hopefully, I have not diluted or skewed one in favor of the other.

I see truth in both science and the Bible and I hope to be able to convey that truth to you in this presentation. Regardless of your position on this topic, your own value as a person does not diminish in my eyes. Believe that, please.

Being able to separate a person from their views has served me well in my dealings with cultists (*as one example*), who are not the ideologies they have succumbed to. Of course, to see faith in this way it helps to distinguish between the scientists and the scientists' faith or ideas. I do not see an evolutionist the same way as I see evolution. To me, the theory is separate from the one who theorizes it. People instantly and inherently deserve my respect for being souls made in the image of the Living God; their concepts, notions, and opinions do not.

You could say that I am tough on concepts, but easy on people. And I am very hard on "evil-you-shun" (*evolution*) and "deep time" (*the millions of years fantasy*). Do I hate

1 | Competing Worldviews

teachers? No, not at all! But I hate what is taught, if it is untrue. I'm not really fond of cults though.

> Be tough on concepts; but easy on people.
> ~ Me

Warning: I'm about to get tough on schools.

While school is not a concept, it's an institution of man and demons—an earthly power. According to the Bible, that makes it fair game as a target in this war of ideologies. School employs people, but is not a person itself. It is a construct; a system; a method, that has become a monster. In your mind, separate the people from the entity of the school. It has likely been there for longer than the staff has been alive. It has a personality of its own, apart from the staff's individual personalities (*to a large degree*). If I come off as mean here, please know that it is not against fallible people, who can easily be ensnared by harmful ideologies without even knowing it. I speak against notions, ideas, philosophies, and theories, as well as systems, institutions, and what the Bible calls "powers"—not against individual people. I speak out against the godless lies of Satan and his evil empire. And while I can still love a person, I can hate their ideas, words, and practices.

But yes, in matters of God, we are always going to run into a point where faith is all we have to continue on with in our search for the big answers (*where science cannot go*). No matter the path that we take, we will all be faced with a divide that requires faith to cross over, that is, if we wish to continue on our journey into God's loving presence. And if you were to ask me, I'd say that public school is a cult; pushing fiction for power. Shall I repeat that?

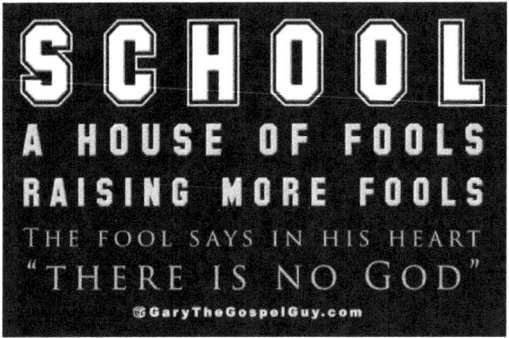

Public school is a cult; pulling kids away from God with a massive network of lies.
~ *me again*

I see it as controlling and manipulative, and at least passively aggressive when confronted—it's sometimes litigious, and it can wield local law makers and enforcement like a rag doll too.

1 | Competing Worldviews

SCHOOL
- S: arta'ns
- C: rauze
- H: ousecur
- O: bsr
- O: uro
- L: ord

Taking the Creator out of the picture of the creation is like taking the coding out of the DNA in our bodies; it's just a jumbled mess and then it dies and begins to stink. Taking God out of the narrative is a lot easier than replacing Him, BTW. *You can push Him out; but you can't replace Him with anything worth having.* The farther away you go from Him, the worse things will get for you. There really is no substitute for having the Living God in your life. I am not against the idea of school; just the ideas pushed by school. And since we pushed Him out of the schools in the 1960s they have only gone downhill—quite a bit.

But this can all be turned around by inviting God and His Bible back into our schools. All we need to do is recognize how bad things have gotten in school since removing God and bring Him back in. And don't cite "separation of church and state" as an excuse. That phrase is not in the Constitution of the United States. What we do have in this country that is established on the laws and Word of God is this: "One nation, under God". We should keep it that way for our own good and success.

This Broken Planet

He said: "Bro, gimme your spear for a sec. This dumb ape keeps touching my butt".
(*Parody of The March of Progress, 1965, original artist Rudolph Zallinger*)

The theoretical artistic license shown in the *March of Progress* (*parody shown above and below*) was extremely damaging to the faith of millions of people in the 1960's and following, about the time when God was removed from school. Zallinger showed apes and men walking in a line as if they were in a progression of evolution. It had no basis in reality but its effectiveness as a propaganda tool was immense. And some liked how it looked on their dormitory wall, I guess (*it was a colorful, multipage graphic that could be removed from the publication it was in*).

> For we are not fighting against flesh-and-blood enemies, but against evil rulers and authorities of the unseen world, against **mighty powers in this dark**

1 | Competing Worldviews

world, and against evil spirits in the heavenly places. (Ephesians 6:12, NLT)

We are fighting using worldviews: demonic versus godly. The public school systems around the world are potentially evil institutions that are taking our children away from our God and giving them to Satan and the powers of darkness. Because the school system is a mighty power in this dark world, it is fair game for us to attack, according to the passage above. I do not mean physically, or violently, or the people personally, or in a manner that violates their human or national rights (*dignity is up for grabs*). I mean spiritually, communally, and politically. Why stand for their nonsense which destroys our moral fiber and society?

Bro said: "Dinner is planned for tonight, Ted"
Then Ted says, "I want those monkeys that stole our clothes!"
(*My wife didn't get it either.*)

Group Discussion Questions:

1. What is Cosmology?
2. Does each cosmology have its own evidence?
3. How do we know that a cosmological view is accurate?
4. Isn't it unchristian to make fun of other views?

Answers:

1. The study of the beginning of the cosmos and life.
2. No. There is only one Earth that we may observe to devise our cosmologies by.
3. The more it fits with what is observed and revealed by the Creator, the more accurate it will be.
4. No. It is unchristian to make fun of other people. We can separate the person from their views. This is in opposition to "identity politics", which tries to judge people based on their views.

Chapter 2 | The Broken Planet Model

> Jesus replied, "I have already told you, and you don't believe me. The proof is the work I do in my Father's name. But you don't believe me because you are not my sheep. **My sheep listen to my voice**; I know them, and they follow me. I give them eternal life, and they will never perish. No one can snatch them away from me, for my Father has given them to me, and he is more powerful than anyone else. No one can snatch them from the Father's hand. The Father and I are one."
> (John 10:25-30, NLT)

Quickly, I just want to let you know that there are times when I don't know if God told me something or not (*speaking of day-to-day things; not visions*). But there are other times when there is no doubt at all (*like with a vision*). This is the second vision of substance I've received in my life from God. The first was in 2019 and was more private. That one was very big and grandiose, whereas this one was small and not as dramatic. This one was very brief; lasting only about five to eight seconds in all; the other one lasted maybe twenty to thirty seconds. The other one was just an overlay of visual spiritual elements over physical reality. This one started like that but then took up my whole view. I instantly understood the meaning and ramifications of this latest vision. As I have progressed through these studies, the Lord has shown me great things. Communicating those concepts is now the challenge before me. Any other visual gifts from God have been fleeting and insignificant, by and large.

Now, don't let my discussing God throw you off of the scientific input that is about to come. Just because God is involved does not exclude science. Not in the least.

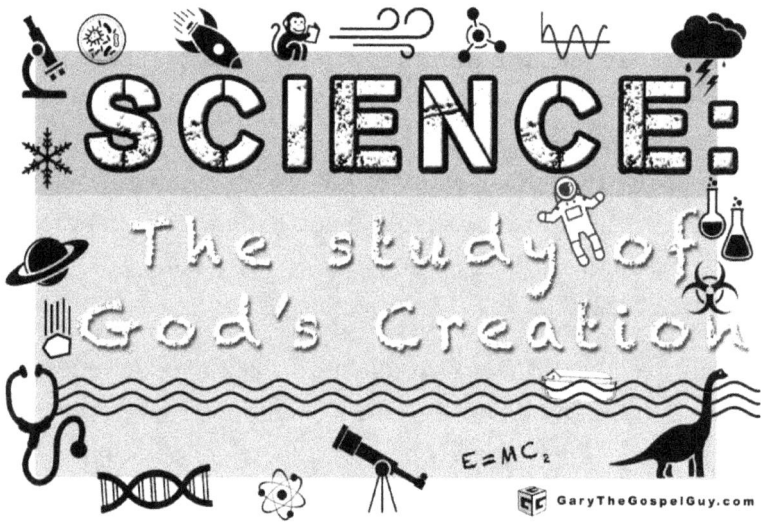

Science The study of God's Creation

In fact, if this revelation is from God, then it will fit exactly with the science (*the real science*) that we gather in the physical side of things today—but not secular scientists' interpretations and backstory. I'm not going to speak in biblical terms exclusively, nor scientific. I like to keep things

2 | The Broken Planet Model

grounded to what we can see and touch if there is a connection, while staying connected to God. That's being scientific in a way. And I'm not planning on quoting a lot of scripture either (*how much is a lot?*). But both lenses are going to be used to look at this interesting topic. And the theory must fit with the Bible or it must be jettisoned as viable. That's how accurate the Bible is. But the Bible must also fit with what we see. And it's up to us to understand what we're looking at. So far, when I take an honest look at the cosmos (*or just life*), the Bible always agrees with the reality I perceive that is on, in, under, and above the ground.

Even though I have no collegiate schooling in the sciences or biblical doctrines of this topic and its related themes, this topic can get pretty technical and "heady". So hang on for a wild ride through the planet beneath our feet and the Maker above our heads. I'll try to explain the technical jargon as we go along, so we're all on the same page as to definitions.

Okay, so now without further delay, I (*finally*) present the understanding that God gave me about Earth's beginning, present, and fiery end. This will actually go by pretty quickly, so buckle in. I give you **The Broken Planet Model**.

"ta da"

The Mantle

Its Irreducible Complexity Reduced

I begin with the Mantle, because it's a key element to Earth's ruin and reshaping. Without having a Mantle that is made like this one, we wouldn't have the conditions that we do on Earth now. So I think it's very important to understand this vision of the Mantle and how it operated while still at peak performance.

The Mantle is actually (*I've learned*) a very complex structure that has (*or had, when whole*) many facets to consider. Just as biologists learned that cells have amazing structure and complexity as technology allowed more discovery into the infinitesimal structures, so too, we find that the Mantle is incredibly complex—more so than I expected, for sure. I'm actually amazed at this work of wonder.

In the image presented to my mind by the Holy Spirit, I saw the Mantle of Earth in perfect condition and new (*not destroyed like it is*

2 | The Broken Planet Model

now). I saw a brown, smooth, rock surface covered by a thick layer of water. The illustration here (*above*) is how I try to explain what I know to be true. Unfortunately, my graphic arts skills have much to be desired. The rocky Mantle in the image is supposed to show where this brown rocky surface was. The vision from God did not show me what lay below or how thick the surficial rock was. But I will show (*below*) two images of geophysical models that depict cool material mixed in with the extremely hot material in the lower regions of the Mantle and Core. This tells me that the smooth brown surface of the Mantle that I saw was blown inward into the Mantle's interior. That, I believe, is what scientists have discovered deep in the Earth's belly. Well, and accompanying the skin of the Mantle, there are probably pieces of the Crust's bedrock that have fallen into it as well.

I have determined, apart from the vision, that the Mantle was a huge rock ball (*almost the size of the whole planet today*), with a firm outer crust and a mushy interior made of lava, magma, and possibly crystalline shapes. It was a perfect sphere; not bulging out at the equator, like This Broken Planet does now. But I only saw its surface; not the interior. The illustrations and descriptions I offer on the interior of the Mantle are not a part of the vision. However, I believe that the Lord has led me through my studies this last year and made it clear to me what actually happened at the Flood. Certain aspects of the Mantle are crucial to understanding this scenario that I will be explaining.

This Broken Planet

In the center of the Mantle is the solid Core, swimming in liquid lava (*they call lava the Outer Core and the solid metal ball the Inner Core*). The Mantle's interior may have been like a gradient of crystalline materials that morph as they go deeper toward the core (*maybe a third of the depth?*). I expect all kinds of states of matter to be in it, which change with depth, heat, pressure, and radiation, and lava to fill in the rest of the space (*until it was destroyed by the blast, anyway, releasing much of the lava*).

So, in your mind, see a gigantic spherical geode-like ball that is filled with lava and has a smaller solid metal ball in the middle (*the size of Pluto; smaller than the Moon*). The exterior surface of the Mantle in the vision was solid rock that was smooth and hard, like a bowling ball. It was a perfect sphere of amazing complexity and power (*nuclear power*) when new. This is the way that I saw it in my mind, smooth and intact. This Mantle had some very crucial functions to perform. It had to contain everything that was inside of it (*especially the heat and radiation*). **Note**: Geodes are formed as bubbles in lava; born of heat and pressure.

It was made irreducibly complex. Meaning, all of the parts had to be in place together for it to work right. It was a work of engineering on a planetary scale. Just like, yet unlike, all of the other planets that were made by the same Guy. (*And don't forget those infinitesimal systems and spheres deep inside the lifeforms with the same kind of physics and feel, too.*)

2 | The Broken Planet Model

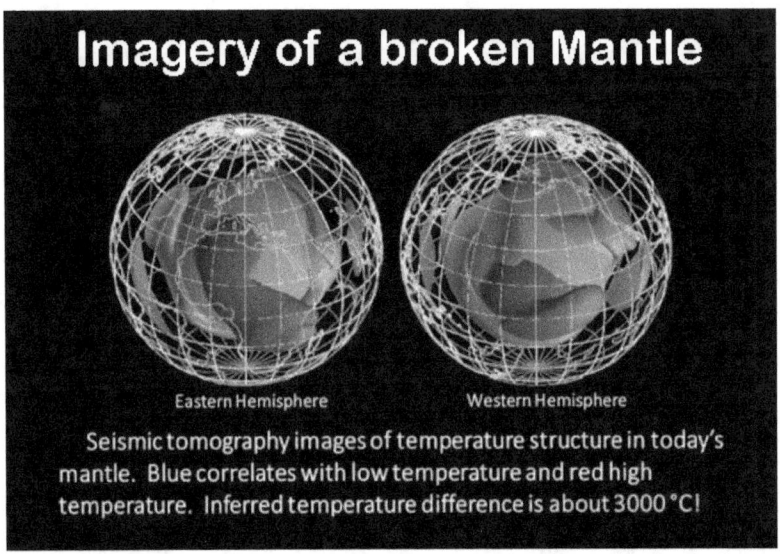

Seismic tomography images of temperature structure in today's mantle. Blue correlates with low temperature and red high temperature. Inferred temperature difference is about 3000 °C!

According to the Broken Planet Model (BPM), the lighter matter shown is what was blown into the inner parts of the Mantle in the explosion. The darker material is the original matter. The image is from Dr. John Baumgardner and the CPT model. The lighter material is either or both the skin from the Mantle and/or the bedrock bottom of the Crust Layer. Both would be cooler than the original inner matter. Having the skin blown inward would allow the bedrock to fall in as well.

Heat

Again, inside of the Mantle was the Core, which is as hot as the Sun's surface (*maybe a touch hotter, even*). That's a lot of heat to contain right under our feet; estimates are around 6,000° C or 10,832° F, but it contained it (*and still does to a degree—pun*). I believe that the crystalline structures are key to containment of the heat. (*When I say that it contained the heat, realize that I mean the majority of the heat, like the first*

10,000° F or so. It still gave off a lot of heat to warm the Crust of the planet—idk what temperature.)

Job one; contain the Core's heat. That much heat released suddenly would not be great for the world. Trust me, before this is over you'll see that our current view of "global warming" is cute and funny compared to the real meltdown that has already begun, and would actually consume This Broken Planet in short order, "geologically speaking" (*ha ha*).

Just this one factor—diminished heat containment from the Mantle—would spell the doom of Earth in the not-so-distant future. (*Hint: not "global warming—global melting".*)

Another possibility is that the Core is now cooling down, since its heat is no longer being retained. And that, too, would spell the end in a distant possible future that will never arrive.

Neither of these two things will actually happen; they are only geological inevitabilities that would eventually destroy us if given time. However, the Lord will return before Greenland is green, so don't worry.

Radiation

The Mantle's second job was to contain the Core's radiation. I don't have any knowledge or data on the *amount* of radiation that was present in the Core in its normal state. But it had radiation that it kept safely tucked inside of it.

Unfortunately, the Crust and sky have quite a bit of radiation in them now. I'd guess it's enough to keep people,

2 | The Broken Planet Model

plants, and animals from living very long lifetimes, like those recorded in the Bible before the Flood.

The introduction of radiation upon the Crust is what caused an almost 100% reduction in the lifespans of people on Earth. Did you ever wonder how Noah and Adam and many of the people in the Bible born before the Flood lived into their 900s (*nine-hundreds*)? This is a theory that explains it. Our maximum age has been held to 120 years (Genesis 6:3).

We may not think that the radiation on Earth is all that bad; but in reality, it's keeping us from living really long lives. These days, a life that long seems impossible. But to them, they would wonder why we are all dying off so young. If I lived back then, at 62 years of age (*my age now*), I'd have been like a teenager of today.

"Did you hear about Uz?"

"Oh yeah, what a shame to die so young."

"Yeah, he was just hitting his four-hundreds, poor guy!"

So, a key thing to remember about this model is that when the Mantle broke open and the radioactive lava spewed out, the world became irradiated for the first time. Rather than long periods of a dribbling of radiation onto the planet, realize that it was a huge dump of it all at once, with dribs and drabs continuing after that.

> A huge amount of radioactive lava coming from deep within the Earth all at once is why radiometric dating doesn't work. ~ me

Is this inconceivable to you? Do you really want millions of years that don't exist? I think this model is totally plausible; you should too. I'm just glad that I don't have to live in this dying, degrading skin for that long.

God intended for people to live forever. It was our introduction of corruption into our makeup that caused Him to put a limit on our lifespan (*by way of radiation at the Flood*). If He didn't introduce death before the Flood, Adam and Eve would still be alive today. If that happened, this world would've been overrun a long time ago with crazy, immoral people who care nothing for others. It would be a MUCH bigger hell than it is now.

Pressure

The third function of the Mantle would have been to contain its own inner pressure (*if any*), which could have been immense—just as the heat within it was immense. As we will see, it was this proposed internal pressure that likely helped the Mantle and the Crust break into pieces. Or, break more quickly and completely; really ruining the Mantle and Crust.

However, internal pressure is not necessary for the model; the ruin could have come whether the pressure were at zero in the Mantle or not initially. The thing that is necessary is at least one hole in the Mantle (*for lava to exit through*). And if that hole were made under pressure...[big, big boom].

And while I'm saying that pressure is not necessary for the model to be viable, I believe that it was present in a great amount. It was the pressure that pushed out the lava that

2 | The Broken Planet Model

formed the foundations of the continents and thereby saved us from having a water world. Lava gushing up from below is a big factor in our geo history. If the pressure were very great, it could have even produced asteroids and comets from Earth debris leaving the planet's gravitational grip (*maybe*).

Division

The fourth job of the Mantle was to keep its interior safe from outside elements that could damage the Mantle if introduced to its contents. In fact, it would have been extremely important to keep what I'm going to present next (a *Water Layer*) away from what is inside the Mantle (*the hot lava*).

"Adam!" He yelled, "Keep the water away from the lava! It'll explode if they touch!"

Whether we look at keeping what's inside away from what's outside, or what's outside away from what's inside, having these two sections exposed to each other turned out to be a very bad thing. Explosively catastrophic!

Have you ever seen water contact boiling oil, or something else that is super hot? It explodes into vapor. You see, when water instantly (*violently*) turns into steam or vapor, it's called an explosion. And when a watery planet has an interior that's as hot as the Sun, allowing the super-hot part to contact the water part is not going to go well. The resulting explosion must have been phenomenal, even though the water was already hot to our touch.

And since the Water Layer was encased in rock (*the Bedrock layer*), making it sealed, the rapid, explosive buildup of pressure would have really let loose when the Bedrock Layer failed. And since the Crust (both Soil and Bedrock Layers) already had a (*presumed*) hole from a meteorite, it was compromised and wouldn't have had the strength to hold together. In your mind, see the bedrock expanding to the point of failure from the inner expansion of water into gas. As soon as that initial breakage of the Bedrock occurred, it began a chain reaction collapse that quickly spread around the world.

So when the two elements (*lava and water*) met, the spark of catastrophe was ignited. The fuse was lit; the powder keg blew. Allegedly, a meteorite moving at great speed punched through the Earth's Atmosphere, Soil, Bedrock, and Water Layer, and then came to rest deep within the Mantle's flesh. This injury brought lava, like blood, shooting up and out from deep within the Earth's Outer Core. Ironically, it was this bleeding that saved the human race. Earth's bleeding lava saved us from being on a water world. That escaping lava came out and made its way onto the surface of the fallen pieces of the Bedrock and formed scabs mixed with the soil skin (*turned mud*) to become raised up scars, places for refuge from the waves. We call these scabs "continents" and "islands" today. The lava both ruined and saved the planet, simultaneously. Without the pressure, heat and escaping lava, the planet wouldn't have blown up; without the pressure, heat and escaping lava the continents would not have been raised and created. The pressure that was certainly created was the

2 | The Broken Planet Model

pressure from the contact of lava and water in a closed sphere, creating expanded water vapor; not what was present inside the Mantle before the contact. But both are likely.

Magnetism

A less obvious fifth job of the Mantle would have been to keep its Magnetic Field intact. I have a feeling that the magnetism of Earth is intended for more important things than just setting our compasses and GPS devices to. But I've heard that these days it isn't what it used to be. And that it could just shift suddenly, too. I have a lot to learn in this area. All I know about this is from an aeronautical perspective (*as a former pilot/controller*), which isn't much. But I have heard that the motion of the lava in the Outer Core, rotating around the Inner Core, creates a dynamo effect, which is the source of the magnetic field.

I see the Crust experiencing a tremendous earthquake (*for lack of a better term*) as it is expanding from the inner explosion, reverberating and shaking just before, and during, its complete collapse. I also see it turning or spinning on the Water Layer as it breaks up, due to the inertia of the meteoric impact and also the inertia of the resultant blast—especially since it occurred in water (*actually, it's probably still in motion*). The irony of a blast happening in a sealed water container of sorts is that its effect was minimized. And we can surmise that it happened at the bottom of the Water Layer as opposed to the top. But the resulting waves must have been something to behold. And keep in mind that the water was in *full motion* <u>before</u> the explosion. (*The Water Layer was in motion, while the Crust Bedrock upon it was not, because the Crust was being held by gravity to the Mantle. I don't know what makes the Mantle rotate. The Moon may have played a part in stabilizing the Crust as well.*)

With the bedrock likely spinning to a different orientation to the Mantle as it cracks and breaks apart, and all of the surficial features melting into mud, there's no way to correlate our surface now with that surface back then. No way. I'm trying to describe what happens to a basically hollow rock ball that is blown apart while riding on deep water that it then falls into as a fractured mess. And that ball had loose soil on it that was washed away by the flowing, scalding water. What we see today on the surface is nothing like it was in the days before this catastrophe. Not one thing on the surface remained on the surface; not one part of the former surface remains as part of the surface now. Every square inch of the antediluvian surface is now under the oceans or continents. Every square inch of our current surface was created or formed during or slightly after the flood, 4300-some-odd-years ago. What was on top is on the bottom; what was on the bottom now makes up the top (*when mixed with some of it*). Soil is the only surviving part of the land; and most of it was turned to solid rock by the rising lava from the depths below. The Bedrock became our current Tectonic Plates.

Wherever North, South, East, and West were before the world exploded, they aren't there now. Did things shift? Oh, yes; undoubtedly. Could they shift again? Why not? I doubt that a broken magnet is as good as one that's intact (*as a*

2 | The Broken Planet Model

metaphor). But this shifting is also happening to the part that is now covered up by the brand new continents; the underlying Mantle. The really important shifting to the magnetosphere is probably a result of what happened to the Mantle. The Mantle must have shifted some when it exploded (*imploded*) inward as the Crust exploded outward. Maybe this touches on the variation between true North vs Magnetic North? Maybe pieces of both the Bedrock and Mantle surface are interfering with the dynamo effect of the Outer Core? This would be a good study for someone to undertake.

Now, I don't think it is the Mantle that produces Earth's gravity; I believe it's primarily the Core (*although this is still up for debate. Newton said all bodies attract each other in space*). If the Core is the center of Earths' gravity (*the magnet*), then having the shell of the "magnet" (*the Mantle*) burst inwardly may have had an influence on the magnetosphere (*the magnetic field around the planet*). On the other hand, if the Mantle is also responsible in part for Earth's magnetosphere, then its demolition would definitely affect its field, I'd think. But maybe it's all academic, because the Mantle includes the Core. But magnetism and gravity are two separate phenomena. I just use these terms for illustration.

This Broken Planet

Another image showing how the Mantle was subjected to a series of explosions when its lava met the Water Layer above blowing parts of the Mantle inward (Image Sebastian Noe ETH Zurich—grayscale does it no justice.)

Magnetic Fluctuations in the Crust

I would also expect smaller pieces of the Crust's bedrock to be upside down in some measure across the planet. Maybe near the margins of plates (*a plate is a big piece*)? Is this another explanation for apparent polarity shifts in the rock in various places worldwide? Are some pieces (*here and there*) just upside down or sideways, and that's why their polarity is different than other regions? That's just something to think about (*there will be more on this in just a minute*). Pieces of all sizes and shapes were flying everywhere in this massive explosion. Some are extremely huge in surface area, while others are tiny pieces tossed about. I'm sure there are all sizes of medium as well. When a glass breaks on the floor, there

2 | The Broken Planet Model

will be large pieces, medium pieces, and extremely tiny pieces, all the way down to dust. Why would this be any different? The measurable polarity shows it.

Another likely reason for apparent shifts in magnetism in the rocks, which is more abundant, could be due to the soil turning to rock as it is still malleable and not yet fully hardened, but still receiving the magnetic signature imprint. While still being hardened, it could be pushed about, causing built-in fluctuations in the magnetic signature as it wrinkles and fractures from the motion.

Tectonics and Subduction

Look at the Tectonic Plates we have today; those are the pieces of the Crust Bedrock that fell onto the fractured and damaged Mantle. Some are on top of others; some got submerged into the Mantle's hot lava depths. This is where the idea of subduction comes from. But it was really caused by the explosion and cracking of a collective of pieces (*originally part of an intact sphere*) that are larger together than the surface they fell onto. Naturally, physics demands that those pieces of a larger sphere must overlap each other, unless turned to small enough bits, when falling onto a smaller sphere. In our case, the pieces were mostly very large and overlap each other in places. However, there is evidence (*shown in the images above of a "Broken Mantle"*) that cooler pieces (*pieces that don't belong there*) are found deep in the much hotter substances below. This says to me that some pieces did in fact go deep into the hot goo; pieces that didn't start off there. (*What say you?*)

This Broken Planet

All of this is regarding the Bedrock pieces now on or in the fractured Mantle; but we can also observe the new rock formed from the hardening mud layers (*that is, the soil that was atop the now shattered bedrock*). When mud turns to sedimentary rock (*through hardening by heat*), a magnetic signature is imprinted into the layer. If it gets subsequently deformed or mangled somehow (*depending on how soft and pliable or hard and brittle the forming rock is*), that signature is already set and we can tell if it was turned. Some rocks are wrinkled up like cow patties and the waves of differing magnetism polarity can be read in their folds. So, to really confuse observers, there are those two kinds of rock: Broken Bedrock pieces (*now tectonic plates*) and cooked mud that hardened into sedimentary rock (*now forming the continents above the plates*). And since those pieces (*the smaller ones*) either flipped or sank into the Mantle, that would show up as anomalies in the magnetic signatures. And since some of them are overlapping, that is what we call "subduction".

2 | The Broken Planet Model

The Water Layer

This Water Layer is something that I don't think you've heard about before. It's a key element of the Broken Planet Model. I see no reason why this Water Layer cannot geologically explain nearly all of the major features of Earth's surface today after the great catastrophe that ripped it apart, and also fit with the biblical narrative, which demands adherence (*due to its veracity in everything else*). This is opposed to the mainstream academic view, which fits with neither scientific observation nor the Bible (*despite their claims that it does*). We can now dispense with that idea and hold onto something that actually works in reality (*Biblical Young Earth Creationism and the Broken Planet Model*).

Depth

No one ever saw the Water Layer as it was. It was hidden deep under the surface of the Earth. It existed a few miles below the surface, under everyone's feet, and they wouldn't have even known it was there.

My proposed Water Layer may have been less than two miles thick and directly covering the Mantle. I don't know precisely how deep it was, since that detail wasn't given to me in the image I saw. I came up with two miles by just looking at Google Earth and reading what the depths of the ocean are (*I mean, it can only go as far down as the Mantle, right?*). I think of the Mediterranean Sea as being about the same average depth as the Water Layer.

According to the National Oceanic and Atmospheric Administration (*NOAA—pronounced "Noah"*), the average ocean depth is 12,100 feet, or 3,688 meters. The deepest parts go down as much as 35,876 feet, or 10,935 meters. So, on land, if you can see an aircraft's contrail high in the sky but the airliner is too small to see very well, that's about the same distance as being at the deepest bottom of the ocean and trying to see a boat on the surface. It's deep. Really deep. But, I think the average depth today is very close to what the Water Layer's depth may have been.

One of water's many amazing properties is that it never really goes away from the Earth. It only gets moved around and changes states (*solid, liquid, or gas*).

Supposing that there could not have been more water on planet Earth than what we have now, I think we could try to estimate how deep the Water Layer may have been, but that would seem to be difficult (*to me*) with all of the variables involved. No doubt, some smart college boys could figure it out.

2 | The Broken Planet Model

First, we don't know how big the single land mass then may have been exactly (*not a continent, but a continuous layer*). Why don't we know this? We'll look at that in the next section on the Crust Layer. But the way to go about determining the depth of the Water Layer involves knowing how much water would have been needed on the surface for life. If the entire Crust were inhabitable, meaning no oceans, that has an impact on how much water is left over for the Water Layer since they were both mixed together at the flood. If it were mostly ocean, well, you get the idea. We have no way of knowing, so we just place *no ocean* as a baseline in the span of possibilities and go from there. (*Personally, I think that the original soil layer was fairly thin and spread out with not too much water upon it or in it.*) This really is not what the water was like though, since there were many life forms in the seas. Some of them were huge, like whales or aquatic dinosaurs. And we know that there were rivers, as mentioned in Genesis.

I would like to ask a hydrologist how much water she thinks would have been needed for life above the surface, if the planet only had land on the surface (*a slightly larger surface than now*). Once we know how much was needed, we can then just subtract that amount from the total amount of water on Earth and then we would know approximately how much was left over for the Water Layer as a maximum. (*Some of you could probably do this off the top of your head.*) But there were lakes, seas and rivers, so good luck.

Function

This Water Layer would have been extremely important to Earth's composition. Being directly on top of the Mantle, it would have acted as an insulator and even-distributor of the Core's heat to the Crust above (*I'll get to the Crust next section*).

The Mantle and the Water Layer together would have been a perfect heater for the new planet's Crust Layer above.

The Water Layer would have been fresh water, without any lifeforms in it, with the possible exception of microbes. (*idk, in my experience, boiling kills germs—microbes.*) It would have been anaerobic (*without air*) as its Mean Sea Level was probably in the midst of the lower Bedrock bottom of the Crust above it.

There were likely currents in the Water Layer from the Coriolis Force, but since there's no shore to crash into, the flow would've been smooth and steady—nonstop.

I wouldn't doubt that the Sun or Moon had some gravitational effect on the Water Layer, but I don't know what it would have been. The gravitational pulls from farther away and acting upon only one side of Earth at a time would be too minimal for me to think they mattered at all, really, versus Earth's gravity pulling from within upon all sides at once and so close, (*yes, even for the Sun—our orbit is the reaction to Sun's pull, after all*). And then, with the Water Layer theoretically in motion, it would be subject to circular forces that stabilize its position within the rocks on both sides of it

2 | The Broken Planet Model

(*containing it*). This means that the water would be less susceptible to being pulled to one side of the Mantle or the other by either the Sun or the Moon (*or Jupiter or anything else, really*) as the water was already going from one side of the Mantle (*Earth*) to the other (*very quickly, at that*). The rotating force of the water cannot be easily overridden by even an equal force, much less by the weak tug of gravity from another body. Completely contained, the Water Layer is very stable as a single mass of energy in constant motion. Like gravity, it has a force of its own that counteracts gravity from afar (*in this sense*). It is so busy going sideways, that it doesn't want to go any other way. (*Look up Angular Momentum, if not familiar.*)

If the Moon had any effect at all on the Water Layer, it may have been to pull some water onto the surface as it passed by that region to water the surface (*since there was water transfer between Water Layer and Surface*). But this is speculation on my part. I hope that someone smarter than me will look into this.

So, having the two spheres arranged this way (*Mantle and Bedrock*), with the one gravitationally pulling from inside the other, uniformly on all surfaces of the outer sphere all at once, ensures that the outer sphere remains stable in its distance from the inner sphere pulling upon it. It would be nearly impossible for the outer sphere to make contact with the inner sphere for this reason. But add in the volume and density of the liquid water in motion between them, and "voila!", a perfect planet for comfort and stability. The liquid Water Layer had mass, volume, density, and angular momentum, as well as intense heat and pressure. It isn't budging in one direction or the other. It is the perfect substance for the role it played then, just as it is now.

Therefore, don't think that the stone ball (*Crust Bedrock*) containing the Water Layer would have fallen into the Water Layer, contacting the Mantle (*as some may think seeing the two stacked up in an illustration*). That's kind of silly, really. It's a ball; that God formed to hold together. How could it fall on its own, without something to make it fall? It had to break to release the water. It might have been strong but brittle at the same time—like most masonry structures, with water seeping or passing through it.

But break, it did, when the layers were punched through and the explosion happened. Still, we know through hydraulics that water is very strong when it cannot be displaced. In fact, it's harder than rock in that regard (*you can crush rock*). If it can't escape the pressure of whatever is on top of it, it just suspends whatever is on it, no matter how heavy. Water has to be displaced for something to fall through it. ***No displacement; no collapse into it.*** This also keeps the Crust from wobbling upon the Mantle, as just discussed. Liquid water can be really dense when captured and unable to escape.

The Water Layer suspended the Crust Layer above it just fine until the Crust was broken, giving the water a path for displacement. And that displacement was the Flood.

2 | The Broken Planet Model

The Flood

> When Noah was 600 years old, on the seventeenth day of the second month, all the **underground waters erupted from the earth**, and the rain fell in mighty torrents from the sky. The rain continued to fall for forty days and forty nights. (Genesis 7:11-12, NLT)

Hopefully, that once cryptic verse isn't so cryptic for you now, is it? "Fountains of the Great Deep" it's called in other versions. "Bursting forth," they say; "underground waters erupted" it says here. Yeah, under a lot of pressure on a global scale at that! Spraying and spewing forth greatly! And when super-hot water meets air, it really goes bananas. I'll get to the rain in a little bit. For now, realize that this hidden Water Layer became the surging waters. The first rainfall was from water being spewed high into the atmosphere from the explosion as it was jetted through the rocky cracks. This initial "rain" is brown and gray from dirt and ash mingling with it. It has pieces of earth joining it in its leap to the heavens and fall back down to earth. It's a cloud of pyroclastic debris and water. It's all on fire from lava and extreme (*sunlike*) heat. Smoke is billowing fiercely as it struggles to keep up with the deadly projectiles in their shooting array of chaos and destruction. The rain from weather phenomena will be discussed a bit later. This rain is fallout. And it's nuclear fallout at that.

So now you know that Earth's ruin at Noah's Flood wasn't just from the rain (*that was not what killed everything and everyone*); it was the bedrock and soil falling out from under them into a deep, torrential bed of scalding, flowing water. They were cooked as they were torn to pieces and drowned. The REAL flood had nothing to do with 40 days of rain covering Mt. Everest by 22'. That would be impossible. (*BTW, the Mariana Trench is about 36,070 feet <u>below</u> sea level, while Mt. Everest is about 29,032 feet <u>above</u> sea level, for a spread of about 65,102 feet, or 19,843 meters, or 12.3 miles. That might be the total thickness of the continents worldwide, from the deepest bottom to the tippy top of the highest heights.*)

    Those waters erupted all right. Some water may have left the atmosphere! Sometimes, *eruption* just isn't a strong enough word. This is one of those times. But the eruption would have been brief, like an explosion. Because it was one. Big earthquake; instant death. Nothing but water for months. Can you see it? Water explodes outwardly from the blast; instantly after that (*simultaneously, really*), swirling, gushing water removes and replaces land; everywhere on Earth; all at once. Kaboom! Flush! Then the world becomes water. But you wouldn't see it from orbit, because of the overcast; from smoke, dust, debris, and fog to the tallest towering cumulonimbus monsters covering every inch of the planet. 1) Green Planet; 2) White Planet; 3) Brown Planet; 4) Blue Planet.

2 | The Broken Planet Model

Green Planet
Covered in lush vegetation without continents.

White Planet
Covered in massive storm clouds from pole to pole and around the globe.

Brown Planet
Covered mostly in mud, with a few pockets of fresh water. Some clouds linger.

Blue Planet
What we have today.

71

The Crust Layer

Finishing out the Layers of Earth is the Crust Layer on top of the Water Layer. The Crust Layer back then was a free-floating ball on a bed of water. It isn't falling through the water for the same reason that canteens don't fall into the water inside of them. Or think of a coconut; it's rounder, like Earth. Best yet, think of a gyrocompass, as a free-floating ball on water. And remember that gravity is holding the Crust and Mantle in a stable state of separation, while the Water Layer provides support to the possibly flimsy Bedrock. But I already covered this above.

Today, we know there is a gap, called an unconformity, that goes by the name of its discoverer: The Mohorovičić discontinuity (*Moho*). This is where the Water Layer was originally contained. Now it's just a place where the two surfaces above and below it are different. It was a gap that held water, but now it's a line that shows a place of separation in the rock strata.

2 | The Broken Planet Model

The Crust Layer would have been the perfect environment for life on Earth—nothing like it is seen today with its current jagged, hostile environs. I see the Crust then as being a hollow ball, filled with water and having a rock mantle in the center that gives off heat. It would have been something to behold.

Composition

The Crust Layer in my model has two basic layers to it: 1) a Bedrock Layer and (2) a Soil Layer on top (*with some surficial water in the mix, too, naturally*).

Picture, if you can, a Bedrock bottom of the Crust (*or look at the pictures*). It's anywhere from 2 to 5 or more miles thick—just guessing. Again, I wasn't given specific numbers, so be loose in your imaginings—or do the math yourself. Actually, the thickness of the Pacific Ocean floor is probably it. For sure, it is whatever lies directly above the Moho.

In the midst of this Bedrock Layer is the Mean Sea Level of Earth's Water Layer that contacts the Crust and probably feeds it with a constant supply of fresh water through springs or vents or a porous bottom. (*Water transfer between Crust and Water Layers is necessary to the model, since Genesis mentions it in the Bible. It also says that Noah was to use tar or pitch to seal the Ark. Hot vents would likely produce tar pits*).

On top of this rock bottom is either soil (*amazing soil full of life that we would not be able to properly imagine today*) or sand (*if aquatic*). Sure, mud, clay, and other soils were there too, as well as metal ores, jewels, and the like (*tar pits too*).

This Broken Planet

The bottom surface of the Bedrock that contacted the Water Layer was super smooth. I imagine the possibility of sand and gravel being at the bottom.

Function

The land produces such diversity and volume and splendor of life that no person today could accurately capture its beauty and vitality. I see nothing but gently rolling hills of various biomes, with who knows what kind of beauty, covering this Crust in one land mass that is slightly larger than our planet is now. I don't call it a "continent" so that it is not confused with the continents of today. Can you imagine walking through a forest the size of the Pacific Ocean? All of where the Atlantic is now was once land. No islands in deep oceans; just freshwater rivers and shallow lakes or seas on an endless land mass everywhere. No deserts or barren lands or deep oceans. All areas are full of life. Antarctica was not there; only more land as warm and fertile as everywhere else on this lush planet. The North Pole was the same—no Arctic Ocean—just lush vegetation teeming with life. No breaks in the land, really; only shallow seas and lakes scattered about with a few rivers here and there, likely in the lowlands.

How about perfect soil that is hundreds of meters to a few kilometers deep and can grow anything in an amazingly short period of time by our standards? How about if this perfect soil were everywhere but under the shallow seas? And what if all the minerals the people needed were easy to recognize and get to and extract?

2 | The Broken Planet Model

Regarding the Bedrock Layer under the soil, this is what we now call the "Tectonic Plates". They are the pieces of rock that used to be in one complete ball, a little larger in diameter than our Crust is now. Some are huge and some are small. Some smaller ones might be upside down. Some are stacked on or overlapping each other (*called "subduction"*). And some may have been pushed into the now mushy Mantle, sunken into its depths. Today, all of Earth is on shaky ground.

Surficial Water

Freshwater lakes and seas could have been abundant and beautiful across the globe. Rivers (*a scar from the massive runoff of the Great Flood today*) may or may not have been as they are now. We do know that there were rivers coming out of the Bible's Eden in the "East". But we don't know if there were any others in the world or not. Most definitely, the Amazon, Congo, Mississippi, and Yangtze were not around before Noah's Flood (*a certainty with my model*). What the rivers were like then is a mystery we'll never know. (*I could see Noah, himself, naming the Tigris and Euphrates rivers as he followed them southward from his landing site toward the Plain of Shinar—and Japheth laughing about it. I mean, if it's the first river to be named, using a familiar name of a large river might be expected.*)

The deepest seas on the Crust Layer would likely have been no more than 1000 feet deep or so (*like America's Great Lakes, or the Caspian or Black Sea, or Utah's Great Salt Lake, and Africa's Lake Victoria, etc.*). The tallest mountains could not have been more than a couple thousand feet tall above sea level. Just gently rolling hills—everywhere, I'd think. Don't think of their seas as being like the deep oceans that we have today. Think only of water like what is on the continents now, at most.

I wouldn't expect a passage from the surficial waters to the underground chambers for anything but water. Only because getting caught down there would be too scary and far too dangerous for any creature. However the water was transferred, it had to have been without allowing animals to pass through. Water that is two miles deep, pitch-black, in violent motion, and hundreds to thousands of degrees hot, and anaerobic is not a friendly environment for any earth life that I can think of.

I imagine the deeper seas to be resting on the top of the Bedrock Layer below. This would maximize the depths and allow for easier transfer of water between the surface water and the Water Layer below. The shallower lakes may have had sandy bottoms. Who knows? But one thing I believe is that the large seas had to be wide enough for the mud to stay far enough away for the sea life to be able to breathe in the waters and to not get scalded to death by the super hot waters below when they rose to mingle together. This would provide pockets of livable temperatures without all of the mud that dominated the planet for a time there. Just guessing.

Water Table

When talking about the water level underground, we call it the "water table"; like the local sea level under the surface. I

2 | The Broken Planet Model

think that the water table would change as the Moon went by. Like now, how it causes tides in the ocean, I think that it could have pulled water up out of the Water Layer onto the surface. Just a hunch, at this point. We need to know what the tectonic plates are made of to get a better idea. There had to be a system of water exchange between the water above and below the Bedrock. That makes the Bedrock porous or having shafts, tunnels, or vents necessary. Maybe it was both (*porous and vented*). It would be neat to know, but not crucial to the model whether it was porous or vented.

Other Features

Nonexistent (*yet*) were the tall mountains, deep valleys, deserts, caverns, and volcanoes that we have on the surface today. Not one tall mountain existed on the whole surface of Earth; no Grand Canyons either. The Crust wasn't as smooth as the Mantle, but it was very smooth compared to today's broken, jagged Crust with soaring peaks and plunging canyons. All of those features are the results of the razing of This Broken Planet of ours.

Pre-Flood Weather

Wherever you would have walked on Earth in that time before the great collapse of the Crust, the weather would have been perfect "birthday suit weather" all the time, regardless of location or date or time of day or night. The reason for the perfect temperature everywhere on Earth is the Water Layer on top of the hot nuclear Mantle, distributing a very even temperature to all places on Earth all of the time.

This Broken Planet

The atmosphere over the Crust would have been very calm too. Today, it's the temperature differences around the world that drive the extreme weather patterns that can be so deadly and destructive to us. Before the Flood, there were no temperature differences great enough to move much wind across the surface at all. Maybe an occasional gentle breeze, especially on the coasts or near vents of steam. I could see that if an ocean heated up too much it would start some local weather, but very mildly. It's difficult to imagine much weather change at all. The air was humid and perhaps more oxygen rich than now.

This means there would have been **none of this:**

- Snow
- Ice
- Sleet or Hail
- High Winds
- Thunderstorms
  - Thunder
  - Lightning
- Tsunamis or Floods
- Earthquakes
- Hurricanes
- Tornadoes
- Uncomfortably Low Temperatures

Features that *may* have existed could have included:
- Blowholes or Pressure Vents
- Springs of water bubbling up from below
- Steam Vents
- Hot Springs
- High humidity
- Tar Pits near hot vents from below (*melting vegetation*)
- Nearly constant atmospheric pressure

2 | The Broken Planet Model

Life on this antediluvian (*pre-flood*) world would have been amazing. Nothing about the weather is trying to kill you—just other men and maybe some animals, plants, and microbes.

I have heard that the atmosphere then may have had a much higher concentration of oxygen, which would have given people and animals a much healthier life. They say you could have hiked forever with that amount of oxygen in the air.

Mid-Flood Weather

Okay, so that was the perfect weather of the antediluvian (*pre-flood*) world. But the Bible talks about there being 40 days of rain worldwide during the Flood. And not just a few days here and then a few days there as it traveled around…no, no, no. I think it means 40 days of solid rainfall, globally, all at once! Which is a lot, to be sure, but it isn't even close to the real damage that was already done with the Crust Bedrock collapsing out from under humanity into the Water Layer. Like pulling the chair out from under someone. Or hitting the bull's-eye on a dunk tank! *FLOOSH!* Face it; God flushed the antediluvian world down the drain—before the rain even began.

And that Water Layer was likely in great motion with much kinetic energy before the explosion. It would be like having the land fall into a gigantic torrential river.

"Instant flood waters just a raging and a raging; coming up from below."

But yes, there was ***rain*** too—a lot of rain. So much rain that the world has never seen that much rain in its turbulent weather systems before or since—not even close.

What causes hurricanes? I've heard it has to do with the ocean water being very warm, making warm, moist air above it. That warm air soon meets with colder air and then it condenses a lot more than it would if it had just come up out of the colder water. (*I hope that I got that right. I'm sure there's more to it than that.*)

So, if ***warm*** water causes huge storms like the ones Florida gets every year, what would happen if the water were ***really hot*** (*like boiling or hotter*) all over the world? And in motion? And what if the air was growing colder because of global sky obscuration (*clouds*) and rain? Because of the explosion, there was an initial burst of water jetted high into the atmosphere that would have taken time to return to the surface as rain, while cooling the already humid air—a blast of clouds and rain to cool the atmosphere.

Well, isn't this the perfect storm, with the perfect ingredients for the most water exchange possible through weather? Some hot water exploded high into the sky all around the globe to fall back as cooling rain. At the same time, scalding water was gushing up through the Crust's Bedrock cracks, as the Bedrock collapsed into the Water Layer, falling onto the ruptured Mantle. The former soft Soil Layer became one with the water and turned to mud. But because the water was so hot as it contacted the air, the weather would have gone

2 | The Broken Planet Model

bananas instantly. Heavy fog and steam would've formed instantly on the surface of the entire globe and then rapidly grown into towering cumulonimbus monsters (*really big storm clouds*) very quickly in a solid blanket over the whole world at once. Really, water that hot would have exploded into the air.

The clouds would have been much bigger, thicker, and deadlier than your worst Kansas thunder-bumpers. They would have been filled with volcanic ash and smoke, each of which is deadly by itself, but this volcanic smoke and ash cloud would have been everywhere over the Earth at the same time, blocking out the Sun and its heat for quite a while (*several weeks? 40 days?*).

Ash would have likely been rare on the planet before this eruption. I would imagine that many of the metric tons of ash on the surface now were delivered in these first 40 days of the flood. But we've had a lot of eruptions since then, and many more to come before it's over. It's amazing the Ark made it at all (*call it God's providence*).

Temperatures worldwide would have gone nuts, while the whole beautiful structure and its efficient ways are dashed and scattered, blended and churned into swirls of destruction. The skies are cooling now from water saturation and blockage of the Sun's needed rays of heat. The sky is quickly super saturated with moisture. Condensation is nearly automatic with the searing seas meeting tepid skies. Cooling from above and warming from below. The world is overcast from sea level to the highest heights of the sky. This is the perfect mother storm of all storms. Earth at this time is a storm planet; a White Planet.

Hurricanes, typhoons, tsunamis, earthquakes, tornadoes, and just high winds… you name it. Also, those seas were already a lot warmer than ours, and the humidity in the atmosphere would have been quite lofty to begin with globally, making the situation ridiculously moist; while the explosion made things ridiculously turbulent. The water is moving so severely that it is moving the already turbulent air above it and vice versa (*everything is in motion from the explosion for a while—all forms of water and "land"*, *which is mud at this time*).

This had to be beyond uncomfortable for the 8 people and thousands of animals locked inside the Ark. I think it was because of the heat that God told Noah to build windows all along the top of the Ark, for life-saving ventilation.

The weather during the Great Flood was the worst weather ever, and will never be so bad again (*as promised by God, with a rainbow to remind us*). 40 days of solid rain actually sounds just about right for all of what was happening at the time, meteorologically.

I invite meteorologists to examine this scenario and talk about what they find. It should be interesting. The winds alone would have been legendary: Tornadoes; hurricanes; El Niño(s) (°.°) …we'll just say it came from, and went it in, all directions.

2 | The Broken Planet Model

So, the weird thing about the rain is that it didn't land on any people. They had all gotten sucked underground (*or water*) and were all dead by the time the rain had started. And the rain started almost immediately (*probably within the first hour, globally*). And if any strong flying birds were able to take to the air, they may have been **drowned mid-flight** before collapsing into the scalding water and mud! (*The air was really saturated with water molecules really quickly.*)

We'll look at Post Flood Weather in the next chapter on the worldwide Flood.

> You should know this, Timothy, that in the last days there will be very difficult times. For people will love only themselves and their money. They will be boastful and proud, scoffing at God, disobedient to their parents, and ungrateful. They will consider nothing sacred. They will be unloving and unforgiving; they will slander others and have no self-control. They will be cruel and hate what is good. They will betray their friends, be reckless, be puffed up with pride, and love pleasure rather than God. They will act religious, but they will reject the power that could make them godly. Stay away from people like that! (2nd Timothy 3:1-5)

And that's just the Church. No, really. Salt and light, Church, Salt and light.

This Broken Planet

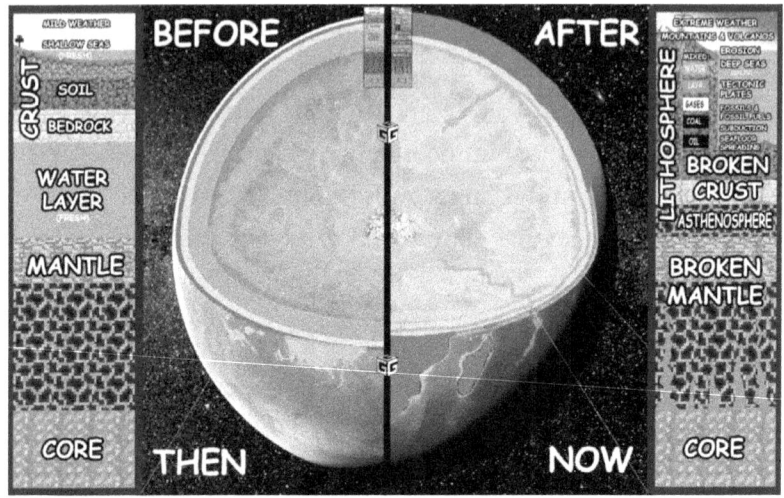

Before and after Noah's Flood

Other Models Fail

Before leaving my model description, I'll just show some verses that eliminate most other models (*creation models*) from consideration due to not matching these Bible passages. Non-creationist models are already eliminated from consideration. This is for the Bible geeks out there. If we are going to say that our model is biblical, then it must fit with ***all*** verses of the Bible. Even these verses:

> The earth is the Lord's, and everything in it. The world and all its people belong to him. For **he laid the earth's foundation on the seas and built it on the ocean depths**. (Psalm 24:1-2, NLT)

Models that have a Pangaea-type continent, like our continents now (*only bigger*), that wasn't actually above the waters, don't quite reach the biblical mark here. The quote

2 | The Broken Planet Model

says that the earth (*dirt*) foundations were ON the waters; not abutting the waters like now. Any model that has huge open oceans cannot be biblical, then. Every biblical model must have ***all*** land on top of water. Yes?

> This is the account of the creation of the heavens and the earth. When the LORD God made the earth and the heavens, neither wild plants nor grains were growing on the earth. For the LORD God had not yet sent rain to water the earth, and there were no people to cultivate the soil. Instead, **springs [or mist] came up from the ground and watered all the land.** (Genesis 2:4-6, NLT)

I know that the Hydroplate Theory (*from Walt Brown*) matches this description, but I don't know of any others, besides my Broken Planet Model. The water table could have been high as well.

> "The Lord formed me [wisdom] from the beginning, before he created anything else. I was appointed in ages past, at the very first, before the earth began. I was born before the oceans were created, **before the springs bubbled forth their waters.** Before the mountains were formed, before the hills, I was born—before he had made the earth and field and the first handfuls of soil. I was there when he established the heavens, when he drew the horizon on the oceans. I was there when he set the clouds above, when he established springs deep in the earth I was there **when he set the limits of the seas, so they would not spread beyond their boundaries.** And when he marked off the earth's

foundations, I was the architect at his side. I was his constant delight, rejoicing always in his presence. And how happy I was with the world he created; how I rejoiced with the human family!" (Proverbs 8:22-31)

Since God set the boundaries for the seas (*post-flood*), which they cannot cross, we have little reason to worry about "global warming" flooding us out. All of Antarctica's snow, if melted, would not add, really, to the vast oceans of Earth at all. And we know that when ice is floating in water, like icebergs, it adds nothing to the water level when it melts into the body of water that it's floating in. (*Look up Displacement.*) This means that the entire Arctic Ocean can melt completely and add not one inch to sea level. The same is true with the Antarctic ice in the southern seas. But Greenland will NOT turn green before the Lord returns. And since it has very little land mass anyway, the snow on the rocks of Greenland is not enough to make any impact at all on sea level. We cannot count the ice in water as adding to sea level; only snow and ice away from the water, on land, can add to its level. And where there is land (*Greenland, the Arctic Circle, and Antarctica*) a water table must be established under and in its soil first. But since these areas are the largest deserts on Earth, they would need to first establish a water table. So that would take care of much of the snow. Makes sense? Let us not panic about "Global Warming", while our souls are in danger of Hell's Fire.

2 | The Broken Planet Model

> He gathered the waters of the seas **like water skins and set the deep in storehouses.**
> (Psalm 33:7, Aramaic Bible in Plain English)

This verse is a difficulty for models that do not contain vast stores of water underground (*without exposure to the atmosphere*). If I understand it correctly, the Catastrophic Plate Tectonics model (*CPT*) does not quite match up with this verse. I think it basically has oceans like today existing before the flood, which were open to the air. That doesn't work, biblically.

> At the blast of your breath, the waters piled up! The surging waters stood straight like a wall; in the heart of the sea the deep waters became hard.
> (Exodus 15:8, NLT)

Quickly, I'll say that the piling up of the waters could refer to the explosion that sent water high into the sky. Indeed, from a distance, water jetting from between two huge plates of bedrock (*with much force*) would resemble a wall momentarily before collapsing to the earth. Into the heart of the sea came mud, which definitely became hard. The Berean Bible puts it like this:

> At the blast of Your nostrils the waters piled up; like a wall the currents stood firm; **the depths congealed in the heart of the sea.**
> (Exodus 15:8, Berean Standard Bible)

These interpretations agree with my model that has soil thickening and hardening in the waters of and after the flood

87

(*the heart of it*). I don't think they mean that the water hardened, as in icing, but rather that there was a hardening that occurred **within** the depths. Here, in the Berean version, the depths are what congealed. What are depths? They don't have to be water, although they are in water. And that's what the model shows; muddy depths congealing in the heart of the sea. Other creationist models do not quite reach this level of agreement with scripture. (*Just saying*.)

2 | The Broken Planet Model

Group Discussion Questions:
1. What is Irreducible Complexity?
2. What do we now call the broken pieces of the former Bedrock Layer?
3. What caused the Mantel to pop, that caused the Bedrock to expand and collapse?
4. Instead of just rain, what was the real destructive force of the Flood?

Answers:
1. The term for something that has to be made in a complete, fully operational state from its creation in order to work. To remove one part of its construction or design would make it unusable.
2. Tectonic Plates.
3. It is believed to have been a meteorite impact. The resulting explosion from lava meeting water blew up the Bedrock Layer above.
4. The collapse of the Bedrock Layer into the Water Layer. It wasn't that the water level rose; it's that the land fell.

Chapter 3 | The Flood

Pre-flood Civilizations

As you will see in this description of the worldwide flood, it is completely impossible for any trace, vestige, sign, hint, remnant, evidence, mark, or record (*other than the Bible*) of human existence to survive it. All life was literally wiped from the surface of the Earth, while the surface was being churned and cooked. Destruction was very quick, merciful, and certain—worldwide.

This goes for any, and I do mean ANY type of structures, buildings, cities, or anything else. No roads scarring the landscape either. Not one tiny little bit of evidence that there were even people on Earth before the flood. Not one stick or stone. Nothing. You'll see why.

The Narrative

If I'm going to go on about the Bible being the best narrative of the world before, during, and after the flood, I should probably share with the reader what that is, shouldn't I? And I'll say a word or two about the text as we go along.

> When everything was ready, the Lord said to Noah, "Go into the boat with all your family, for among all the

> people of the earth, I can see that you alone are righteous. Take with you seven pairs—male and female—of each animal I have approved for eating and for sacrifice, and take one pair of each of the others. Also take seven pairs of every kind of bird. There must be a male and a female in each pair to ensure that all life will survive on the earth after the flood. Seven days from now I will make the rains pour down on the earth. And it will rain for forty days and forty nights, until I have wiped from the earth all the living things I have created."
> So Noah did everything as the Lord commanded him. (Genesis 7:1-5)

Notice that it talks about "kinds" of animals? If the colleges cared about being biblical, they'd use this term. But since they're stubborn about that kind of thing, the closest we can get to the biblical "kind" in biology today is the "family". So if you already have a framework in your mind based on the schools' classifications, just think, "family" when the Bible talks about "kind". The idea is that the animals are breeding pairs that can produce the species that we see today (*or since extinct*). "Dog" and "cat" are examples of "kinds". There are many varieties of each, but all the varieties can mate with each other and produce offspring that can also reproduce. Also, dogs and cats cannot interbreed with each other. A hybrid of cat and dog is not possible, biologically (*only in cartoons*).

Notice too, that God "wiped from the Earth" its previous life. I take this absolutely in a most literal sense. In my model,

3 | The Flood

the ground was removed from under the people and other life on the surface. All plants, birds, animals, insects, and people lost their lives within probably one hour of each other globally, as the water stripped the bedrock bottom of the Crust Layer of all of its soil and life. Yes, they were literally *wiped* from the Earth's Crust Layer, along with its soil.

> Noah was 600 years old when the flood covered the earth. He went on board the boat to escape the flood—he and his wife and his sons and their wives. With them were all the various kinds of animals—those approved for eating and for sacrifice and those that were not—along with all the birds and the small animals that scurry along the ground. They entered the boat in pairs, male and female, just as God had commanded Noah. After seven days, the waters of the flood came and covered the earth. *Genesis 7: 6-10, NLT*)

So we see a grand total of eight people surviving the great breaking of the Earth and cleansing of humanity. Estimates of the Earth's population are in the millions at the time. They took 14 sheep (*for example*) onto the Ark, but only 2 "wolves" (*dog-kind*) and 14 of each bird.

Jewish Religious Law, called "halakha" by Orthodox Jewish adherents, discusses what it calls "clean" and "unclean" animals. Clean animals are those that are approved for food and sacrifice; unclean are those that are not. Some versions of the Bible use these terms in this passage.

Having such a large number of sheep-like (*ovine*) and cow-like (*bovine*) animals alone would have been plenty of food for Noah and his family (*post-flood*). But they also had chickens and other fowl for eggs and meat (*since meat is on the table now*). I'm guessing that the family had a bunch of seeds onboard too—maybe fruiting plants? After all, Noah was a 600-year-old vegetarian. I'll bet he had potted plants on board—maybe even fruiting, at that. I mean, he had to think of more than just what food to take onboard for the journey, but what food to grow once their new life was under way. And seedlings would be appropriate.

The idea that there were only two of each animal is wrong. Clean animals (*livestock for food and sacrifice*), and all birds, were much more numerous (*14 each*) than the unclean (*2 each*).

I find it interesting that Noah knew about these distinctions before the Law of Moses came some 1,073 years later. This tells me that some of the laws of Moses were instituted by God prior to Moses' receipt of the Law, in circa 2,729 *anno mundi* (*AM; year of Earth*). (*According to my studies, Moses was born in 2,649 AM and received the law at age 120.*)

Incest

Before going any further, and since I just mentioned the Law of Moses, I should talk quickly about incest. You know, incest is a brother and sister having children together.

3 | The Flood

One of the more prevalent objections to the Bible's narrative is that it depends upon this practice that it also bans.

"Why did Adam and Eve's children (*however many there were*) have children together?"

"Because they had no one else to conceive with."

"Well then, why did God say that it's against His law later?"

Because the sin that man introduced into their DNA brings with it entropy, which ensures that each generation of DNA information is degraded just a little bit. And the more alike the parents' DNA is to each other, the more rapid or pronounced the breakdown.

In Noah's day, twins (*a boy and a girl*) could have had perfectly normal children together, because their gene pool was just that deep back then, unlike ours now. If I were God (*ha ha*), I'd have made girls in pairs at first, to really move things along. But it was normal to mate with siblings for a very long part of human history (*2.5 millennia*). Only with the Jewish Law did it become forbidden, when it was no longer necessary or beneficial (*becoming a detriment to our health because of genetic decay*).

Backstory

> [Adam and Eve's children] began to multiply on the earth, and daughters were born to them. The sons of God saw the beautiful [daughters] and took any they wanted as their wives. Then the Lord said, "My Spirit

will not put up with humans for such a long time, for they are only mortal flesh. In the future, their normal lifespan will be no more than 120 years."
(Genesis 6:1-3, NLT)

This passage is just packed with controversy.

First, we have the "sons of God" checking out how beautiful the men's wives, daughters, and sisters were. Who are these "sons of God"? Well, according to other ancient documents, scrolls, books, or accounts, they were angels who were charged with overseeing or watching the humans. Their title is often given as "Watchers".

In another place, the Bible talks about them getting a special punishment by being thrown immediately into the Abyss or Tartarus (*prison for bad angelic beings*).

Here's a quick side note:

> For God did not spare even the angels who sinned. He threw them into hell [Tartarus], in gloomy pits of darkness, where they are being held until the day of judgment. And God did not spare the ancient world—except for Noah and the seven others in his family. Noah warned the world of God's righteous judgment. So God protected Noah when he destroyed the world of ungodly people with a vast flood. Later, God condemned the cities of Sodom and Gomorrah and turned them into heaps of ashes. He made them an example of what will happen to ungodly people. But God also rescued Lot [a man] out of Sodom [a city] because he was a righteous

3 | The Flood

man who was sick of the shameful immorality of the wicked people around him. [They were sodomists—practitioners of anal copulation, (especially male on male).]

Yes, Lot was a righteous man who was tormented in his soul by the wickedness he saw and heard day after day. So you see, the Lord knows how to rescue godly people from their trials, even while keeping the wicked under punishment until the day of final judgment. He is especially hard on those who follow their own twisted sexual desire, and who despise authority.
(2 Peter 2:4-10, NLT)

So the story goes that angelic beings had somehow mated with human women, making a kind of angelic/human hybrid super race. God was not happy about this, if true. Either way, it seems that God managed two judgments at once: the angels went to a hellish place, while the Earth and nearly all of humanity and other land-based life was obliterated by Him directly. They could have both been sanctioned simultaneously.

In those days, and for some time after, giant Nephilites lived on the earth, for whenever the sons of God had intercourse with women, they gave birth to children who became the heroes and famous warriors of ancient times.

The Lord observed the extent of human wickedness on the earth, and he saw that everything they thought or imagined was consistently and totally evil. So

> the Lord was sorry he had ever made them and put them on the earth. It broke his heart. And the Lord said, "I will wipe this human race I have created from the face of the earth. Yes, and I will destroy every living thing—all the people, the large animals, the small animals that scurry along the ground, and even the birds of the sky. I am sorry I ever made them." But Noah found favor with the Lord. (Genesis 6:4-8, NLT)

The angelic-human hybrids were called Nephilites, or Nephilim. (*J.R.R. Tolkien's elves may have resembled them perhaps. Others might say that they were more troll-like.*) Some think they were actual giants (*as tall as 30 feet by some claims*). Who knows for sure? Nobody. It's all guesswork when it comes to this topic. And it's a sideline topic at best.

Want a juicy tale? How about this one: The Lord threw the Watchers into the center of the Earth, crashing them right through the Crust and Mantle, breaking and wreaking havoc on Earth, killing almost all life.

It isn't in the Bible and I just made it up, but it's as good of an explanation as any right now. The bottom line is that God killed nearly all humanity and every other life-form that breathes air. He did that because He was not happy with mankind and their violent and sexually perverse behavior.

The Story of Noah

> This is the account of Noah and his family. Noah was a righteous man, the only blameless person living on

3 | The Flood

earth at the time, and he walked in close fellowship with God. Noah was the father of three sons: Shem, Ham, and Japheth.

Now God saw that the earth had become corrupt and was filled with violence. God observed all this corruption in the world, for everyone on earth was corrupt. So God said to Noah, "I have decided to destroy all living creatures, for they have filled the earth with violence. Yes, I will wipe them all out along with the earth!

"Build a large ark ['box'] from Gopher wood and waterproof it with tar, inside and out. Then construct decks and stalls throughout its interior. Make the boat 450 feet long, 75 feet wide, and 45 feet high. Leave an 18-inch opening below the roof all the way around the boat. Put the door on the side, and build three decks inside the boat—lower, middle, and upper.

"Look! I am about to cover the earth with a flood that will destroy every living thing that breathes. Everything on earth [dirt] will die. But I will confirm my covenant with you. So enter the boat—you and your wife and your sons and their wives. Bring a pair of every kind of animal—a male and a female—into the boat with you to keep them alive during the flood. Pairs of every kind of bird, and every kind of animal, and every kind of small animal that scurries along the ground, will come to you to be kept alive. And be sure to take on board enough food for your family and for all the animals."

> So Noah did everything exactly as God had commanded him. (Genesis 6:9-22, NLT)

This tells us that God destroyed Earth because of man's wickedness. It also hints at a problem with race purity, since the fallen angels allegedly got busy with the human girls (*not knowing how that works, exactly*).

God killed every animal that breathes air on planet Earth (*not in the sea*). He killed every man, woman, and child on Earth too—except for those eight and the beasts in their care.

Notice that the animals came to Noah's family. They didn't have to go find or round up anything. I would expect them to have stalls, cages, barns, or fenced-in areas near the Ark to keep them in until ready to board.

This Ark has the same basic proportions as an oil tanker; not the same size, but about the same proportions. It was made to be seaworthy in its shape and construction. It had a roof built on with a small (18-inch) horizontal window, gap, or slit along the wall at the roofline all the way around.

Ark Capacity

If you are wondering if the Ark was big enough to hold so many animals and their food and water, you may rest assured that there was plenty of room in the Ark for all that it held. The measurements given above are using an 18″ cubit, but the actual cubit used by Noah could have been up to 24″, making the Ark even larger.

3 | The Flood

The average size of everything on the Ark is pretty much a sheep size. And since Noah was only loading "kinds" of animals, not all of the different "species", it wasn't such a high number after all. (*It just occurred to me that Darwin's finches could have all been represented in the Ark, since there were 7 pairs of them onboard. And who says there couldn't have been some variety already in many of the pairs received by Noah and crew?*)

They didn't have to take full-sized adults; juveniles would do, since they would be onboard for a year anyway. I'll bet that more mice and rabbits came off than went on—just saying. This is another reason to take the younger animals that aren't at breeding age yet. Otherwise, leave room for growth, I guess. I'd suppose that all animals have the ability to enter into a deep sleep for extended periods of time (*guessing*). Certainly, God could induce such a thing on any being (*not guessing*). However, this doesn't mean that the animals had to be hibernating. God told Noah to take enough food for the animals. That implies being awake. But I doubt if they fed them when it was really rough.

Were dinosaurs on the Ark? Absolutely. Any land animal that breathed air needed to be onboard. The dinos were there. Personally, I'd put the amphibians and other water lovers down in the bottom deck, where water would be more likely to accumulate as it sloshed in through the window and ran down inside to the bilge, under the bottom deck. But that isn't proven.

Maybe the birds were all on the top deck, since they're smaller and lighter? And maybe the important animals—the livestock or "clean" animals—were on the middle level? Have fun and see the Ark in your mind and how the eight people took care of them in their sections. It's all totally plausible. Skeptics are the ones with little to no imagination or ability to do the math. Guys a lot smarter than me have done the math and know that there was more than enough room for food, water, and all.

The Global Flood

> When Noah was 600 years old, on the seventeenth day of the second month, all the underground waters erupted from the earth, and the rain fell in mighty torrents from the sky. The rain continued to fall for forty days and forty nights. (Genesis 7:11-12, NLT)

What caused the breakup? He didn't show me that. One thing that is interesting though is how Iceland is a cap on a shaft that seems to go to the very deepest part of the Mantle. It's like an opening filled with lava that goes deeper than anything else. Maybe that is where a meteorite or other object punched through the Crust and also broke the Mantle (*like a bullet*)? Who knows? One thing is for sure, there would be no crater from the impact, since the impacted Crust fall apart and sank into the Water Layer. It was punched right through and then fell apart when the explosion occurred.

"Underground waters" would be my proposed Water Layer. They erupted from under the Crust Layer because the

3 | The Flood

Crust Layer was broken and pushed the water out as it fell. Saying "the rain fell in mighty torrents" is probably an understatement. There is no way to accurately capture the scale of the storms upon the planet at that time. It had to be an all-over storm. But hang on; I'll get to the rain.

An initial explosion sent water, water vapor or steam, ash, smoke, dirt and debris high into the atmosphere (*or beyond*) to fall back as rain, while the entire earth reeled from a massive earthquake. It shot high into the sky very quickly but fell back to Earth more slowly. But before the drops, rocks, and debris could hit anyone on the head, they were already below ground (*in water and mud*), quite dead.

Example of an eruption a very tiny one compared to the explosion of Earth traveling all around the planet in two directions

It was an explosion that travelled around the globe...followed by everything "land" falling out from under everything upon it, being rapidly replaced by surging, searing,

103

steaming, scalding, suffocating water, becoming sludge, becoming deep oceans on the surface.

Throw in some lightening and deafening thunder, or thunderous noises. Then all goes gray above the churning surface as visibility goes from clear to zero in nothing flat from the smoke, ash, steam, fog, clouds, and rain that quickly cover the entire planet. Even when the winds reach gale forces and beyond, your hands would be invisible to your eyes. All life is gone in minutes worldwide and the rains begin with great fury right away. Rain like the world has never seen since. There was likely little or no rain to speak of before this on Earth. But now the emerging water from the depths is so hot that the rain is practically flying upward in the steam. Indeed, liquid water does fly at first and continues for a bit afterward, especially when the water is super-hot and meeting tepid air. This heat is enormous. No mortal creature, of flesh or plant, of land or air, had any hope of surviving the first 5 minutes of this colossal cataclysm. Birds that could escape the earth falling out from under them could not escape the deadly heat blast, or either drowned or burned midair, or fell to the churning, scalding sea. The atmosphere would have been hot for a short time there, as great heat escapes from under ground and water. But it would have begun to cool almost as rapidly, I believe.

There had to be pockets of livable water for the sea life. The water it was all in likely stayed somewhat together until the conditions allowed the sea life to acclimate to the new

3 | The Flood

water. Also, salinity probably came on slowly enough for the life forms to adapt.

This eruption likely started in one location (*Iceland?*) breaking rifts in the Crust Layer all around the world (*the cracks we see in Google Earth—the Tectonic Plates that everything rides on*). Imagine a blast like this ripping its way around the world, cutting those lines that are now at the bottom of the Ocean floors. They cut southward, and also northward from Iceland (*now*) across the North Pole, then through Asia and the Pacific southward. Then the crack spread from there to everywhere on Earth, very rapidly. As it broke under pressure, the explosion sent dirt, ash, smoke, and water high into the atmosphere to block out the sun's heat for weeks. It rained rocks, dirt, ash, and water from the smoke and clouds.

That isn't the rain that lasted for 40 days though; it's only the first rainfall. The real rain started to fall because the water escaping the Crust was so hot. All of that hot water under a cooler and cooling sky was the perfect combination for a deluge we cannot even imagine today. It only took hours (*if not minutes*) for the storms to grow, and weeks for them to stop, especially with the high humidity worldwide at the time it started.

There is great turbulence caused on Earth's surface from the bedrock, deep under the soil, fracturing and falling over a mile, maybe two, downward into the Water Layer below. Swirling currents in the soil and the air and the water are like giant eggbeaters as everything goes crashing down with much turbulence. The soil is turning into mud in the water. All land life is being blended together into a hot, muddy, glob of boiling goop. Some sea life miraculously survives in pockets of livable water. All sea life was already living on top of the falling bedrock and soil, but it was turbulent and dangerous everywhere during the flood. Much sea life died. Tsunamis of hot mud take entire aquatic biomes by surprise, engulfing them like ants being rolled into dough. The Water Layer it's falling into was likely in quick motion already, with much inertia. Then, add in the explosive blast's impact on all of that water, rock, lava, and soil. Turbulence may have been incalculable. *Take that as a challenge, physicists.*

    Air and water are instantly in motion around the world from the explosion. The world exploded and broke apart, sending pieces of the Crust deep into the water and onto, even into the ruptured Mantle below. Above surface, on the heels of the initial up-blast, comes the downward airflow that soon becomes a sideways airflow at the churning surface, and then magnifies with the updrafts of the heat from the scalding muddy waters of destruction, becoming swirls of water and air; vortices of fury, with winds far above 100 MPH, and water currents almost as fast. The wind and water currents are fierce. Tornadoes and other vortices are going nuts instantly with the sinking of the land and the rising of the storm with what was once the Water Layer. Soon, the only forces checking the churning waves are more waves in opposing motions. Ditto for the raging air currents as well.

3 | The Flood

We kind of know what such turbulent weather looks like on a much smaller and calmer scale. Floridians and others in the world have experienced hurricanes for decades or centuries. We just have never seen it on that scale and intensity before, all over the planet at once. On our current scale of 1-5 it would have been a category 100! And Richter couldn't have made a scale for that earthquake that removed all land, replacing it with hot water.

Now, the Bible doesn't say this, but I just know that a huge earthquake was felt all over the planet when the Crust broke away from the blast zones. The biggest ever, with no equal possible. How could it not? I mean the ground literally fell out from under everyone; gobbling and chewing them up as it was falling apart and sinking! There is no running to higher ground when all of the ground is falling out together. And doing so faster than anyone could get away from it. And it wouldn't matter since it's happening everywhere at once. Literally, there was nowhere to go but down to death's muddy, boiling grip. Birds take flight with nowhere to land, which is just as well, since they'll probably die midair anyway from heat or debris or waves or drowning (*yes, birds drowning as they fly through the water-drenched air*).

> That very day Noah had gone into the boat with his wife and his sons—Shem, Ham, and Japheth—and their wives. With them in the boat were pairs of every kind of animal—domestic and wild, large and small—along with birds of every kind. Two by two they came into the boat, representing every living thing that breathes. A male

and female of each kind entered, just as God had commanded Noah. Then the Lord closed the door behind them. (Genesis 7:13-16, NLT)

I'm glad it says that the Lord closed the door. That would have been a problem for me otherwise. (*Just trying to figure out how they did it—not that it would be impossible—more of a timing and sealing issue.*)

By the time our survivors got over their initial jolt of moving water, alas, everyone else around them (*or across the world*) was dead. And without one single solitary trace of any existence of anything but what was floating in the water. Maybe a hat or a shirt caught in the limbs of a floating tree, but even that is unlikely, since the falling mass of earth was sucking everything above down with it in its wake.

For forty days the floodwaters grew deeper, covering the ground and lifting the boat high above the earth. As the waters rose higher and higher above the ground, the boat floated safely on the surface. Finally, the water covered even the highest mountains on the earth, rising more than twenty-two feet above the highest peaks. (Genesis 7:17-20, NLT)

Water has amazing properties, but reproduction is not one of them. Water is a substance of finite presence. There's only so much water in the world. But then we don't lose any of it either. So the water level could have only grown deeper because the land had fallen out from under it. I mean, if the water level were already inside the land's elevation (*the*

3 | The Flood

bedrock part of the Crust in my model), then that's what was happening—the land falling into the hot Water Layer.

"Giant" sinkhole in Mexico Now imagine ALL of the land falling 2 miles deep into scalding water

Yep, the land fell; raising the level of the water from being below it to being on top. And if the soil hadn't gathered into the continents we have now (*all raised up due to heat hardening it as it collected upon escaping lava*) then we would all be under water. I hope that all readers are getting this; the water came up through the ground. Violently. Suddenly. Irrepressibly. Relentlessly. And it swallowed up all life on the planet—save those lucky few that were saved in the Ark. This is not rainfall reaching up to Mt Everest. It's Mt. Everest being birthed from a raging water planet in the throes of chaos.

So, the reference to the highest mountains being 22' below the crest of the floodwater is interesting. For one thing, the mountains were being formed during *and after* the flood. This means that the highest elevation at the end of the flood's year could have been higher than during the flood itself (*in fact, it had to have been*). The elevation kept changing for a few months there. So this reference is just a snapshot moment in time.

However, that doesn't really matter. Sure, with this model we might be able to figure out how high the waters were during the flood and then see what the terrain might have been doing if we could model such a thing. But that would be very iffy at best. It was all just swirls of mud for a while there.

> All the living things on earth died—birds, domestic animals, wild animals, small animals that scurry along the ground, and all the people. Everything that breathed and lived on dry land died. God wiped out every living thing on the earth—people, livestock, small animals that scurry along the ground, and the birds of the sky. All were destroyed. The only people who survived were Noah and those with him in the boat. And the floodwaters covered the earth for 150 days.
> (*Genesis 7:21-24, NLT*)

Initially, the water would have boiled or burned the people to death as quickly as they were drowning and being ripped to shreds. The only thing that saved the Ark passengers from the heat would have been the rapid cooling of the waters by blending with the cooler soil. I can only imagine that as being the thing that saved our grandparents' (*aunts' and uncles'*) lives onboard. If those temps had persisted for long, I think they would have perished just from the heat. But as physics

3 | The Flood

would have it, adding cooler dirt to scalding water would have a cooling effect. Also, the atmospheric obscuration around the whole planet at once had a cooling effect going forward (*a really big cooling effect, going forward*). Oh, and just the release of it into the atmosphere was huge in cooling it down (*especially since that release produced cooling clouds*).

So after 40 days of the water sloshing around and escaping from under the Crust Layer, it finally was all out of its deep abode and covering the whole of the Earth. And by then the storm of all storms had ended too. But the flood wasn't over yet. There was still land building to do, using volcanism and mud gathering/hardening. Wherever the continents are, shows where the volcanism was greatest and where it captured the transient mud of the former soil layer.

At the 150 days mark, the water started to subside from the fresh continents that were beginning to "peak" out from under their bathwater. This happened in-part due to massive, continent-sized sheets of newly hardening mud (*into rock*) being slammed into each other and getting broken in the process. When these sheets collide, they break and wrinkle up into mountains or valleys or canyons. And the lava that was released in all of this produced many volcanoes, with all of their fun features. The Rockies, Andes, and Himalayas are examples. Because tsunamis were so very extreme, they piled up gigantic masses of hardening mud to form the land continents. When India slammed into China, for example, the Himalayas were formed by the wrinkling of the land. And then the volcanism brings in new minerals for making the land rise too. More lava means more rock, and more heat to grab ahold of and to cook more mud into rock.

When I talk of "volcanism", I mean that lava came up through the cracks of the plates. It came out on top of the plates where we see continents today. That lava was immediately covered by the mud that quickly turned to rock (*sedimentary rock*) and formed the continents. This means that all continents and islands were created in-situ (*where they sit now*). This takes away the notion of a Pangaea or other original continent on Earth. I can show another reason why Africa and South America look like they were once one continent. The explanation you have heard for this may be clever and look right, but it is quite wrong.

In a way, the destruction and salvation of Earth through this flood shows the same kind of fingerprints that can only be produced by the Creator. I mean, just as the creation is irreducibly complex, so too are the conditions required for creating the conditions we now observe on planet Earth. Fossils and fossil fuels in the quantity and variety that we see now could only have happened in this fashion at one time; not over many eons. The same is true with the erosion patterns, tectonic plates, and continental shelves. The Flood of Genesis is the only comprehensive explanation for all geophysical and geological features of This Broken Planet. And it was a one-time inimitable event.

The Bible is quite clear about this being a global flood. Did you see the part that says, *"The only people who survived*

3 | The Flood

were Noah and those with him in the boat"? And when it says that the floodwaters covered even the highest mountains on the earth, that also is quite clearly saying that the entire world was covered by water all at once, killing everything on land that breathed air and was not on the boat. I mean, for a time, all dirt was mixed with the floodwaters (*that had to suck for those with gills*). Mountains are made of dirt being scooped into ripples. At 22' it was probably the highest water level and the level of the submerged land/mud at the time (*but not 29,000 feet*).

> But God remembered Noah and all the wild animals and livestock with him in the boat. He sent a wind to blow across the earth, and the Floodwaters began to recede. The underground waters stopped flowing, and the torrential rains from the sky were stopped. So the Floodwaters gradually receded from the earth. After 150 days, exactly five months from the time the Flood began, the boat came to rest on the mountains of Ararat. Two and a half months later, as the waters continued to go down, other mountain peaks became visible. (Genesis 8:1-5, NLT)

I just realized; God's wrath, when the Lord returns, will last exactly the same timeframe—5 months, or 150 days (*see Revelation 9*). (*Huh.*)

The absolute height of the floodwaters was probably there for a very brief time. Almost as soon as it crested it started to recede, as the continents were forming below and growing from lava flows and mud flows coming together and expanding as they hardened and baked. At 22′, all of the former Water Layer was unleashed and out in the open. But the sludge was in motion and so was the lava from the Mantle below. When and where the two came together, continents arose from the depths of the muddy waters. Lava forms the foundation and baking; hardening mud forms the rest.

Earth's foundation was definitely still in motion in places, due to the new Tectonic Plates being formed for the first time and finding their relative positions with jostling, slipping, and falling happening, and with the sludge still in motion and forming into mountains, canyons, islands, and valleys. Of course, any motion like that is going to produce tsunamis on the surface. I mean, how could it not, with all of the motion of the waters and that slippery, ruptured surface of the now mutilated Mantle. This had to continue for months after the explosion. And it did. And this is how the continents rose out from the waters. Lava bubbles up and turns to rock, expanding as it cools; while the mud gets hung up on it and bakes itself into rock and also expands as it hardens. This is the process that formed the continents—all of them. The mighty winds helped to blow the water off the highest parts of the continents and get things flowing to the sea. That we have any soil at all, able to support life, is a statement to the providence of God. And this is why vegetarianism gave way to eating whatever can sustain life.

There would have been much breaking and moving of the crust bedrock pieces—the new tectonic plates—grinding them down and pushing them all around the world. The water would

3 | The Flood

have pushed them like jigsaw pieces being mixed up in the mud and water and lava. The flows were incalculable for a time there, with great mud globs stretching across huge swaths of the planet, coming together as they melt into one another and harden with the lava rising up from below. We can see where certain events happened just by reading the signs in the ocean floor and other places. Things like a lot of pressure flowing up through (*what we now call*) the Mid-Atlantic Ridge, sending huge swaths of mud in opposite directions (*forming the South America and Africa rift—same with Europe and North America.*)

The continents of Africa and South America were not in one piece and moved by tectonic activity. They were formed as the mud (*and maybe lava*) was pushed in different directions due to the water gushing from under the plates through that crack we now call the "Mid-Atlantic Ridge" (*halfway between them*). The force of that water escaping was tremendous, being fueled by the weight of the falling plates. And since it was from a single source and time, it sent the huge globs in equal distances from the rift. (*Nothing slow or gradual about it.*)

Water currents were so fierce that the Ark could have been torn to shreds if not for God's loving care. Winds of immense magnitude would have been raging on for days and weeks at a time. Imagine lying in the tossing Ark, rocking back and forth, to and fro, up and down, listening to the gale force winds howling through the cubit-high window above, that displays only a dark-grey background with rain pouring in as it pitches to receive or deny the drops' entry. But then, the rain is "falling" in all directions, as the wind takes it along its chaotic rumble in all directions; swirls, really. Lightening only lessens the gloom in its fits and spurts, while thunderclaps are ear-shatteringly near and relentlessly, ceaselessly afar off. You clutch your mate a little more tightly as the darkness rocks you into an uneasy fit of sleep (*counting bleats of terrified sheep*).

So what was once the highest ground on Earth under the flood's crest would have been much higher a little later on, when India slammed into China, making the tallest peaks, for example, or Denver's Rockies coming from where Hawaii is now.

Water ran off of the still rising mountain ranges and the forming valleys and canyons. I think that the Grand Canyon is just a big crack that appeared in the hardening mud/rock layers as the continent was still in flux. And then a lot of runoff rushed through on its way to the Pacific. (*This was a common scenario all over the world.*) East of the Continental Divide, it ran into the Atlantic Ocean or the Gulf of Mexico (*oh, or the Arctic too*).

The final runoff at this time would have been something to behold. Like a tsunami that seems to never stop, or a river more than a thousand miles wide; carrying dead animal parts, all kinds of foliage, and rocks—lots of rocks—out to sea, while the volcanoes spit, flow, and roar with lava and mudflows slashing through everything in their path. Man! Devastating! Catastrophic! Epic! (*More words needed.*) But if

3 | The Flood

the Mantle had not been changed, the water would've just gone back to where it was, I would imagine. Thank God for rising lava and baking mud, working together to build up a continent or few (*ok, seven*).

In the Flood, entire biomes were carried away in mudflows for hundreds or thousands of miles and deposited in their baking land encasements that become the new continents. Do you get it? Before the flood, let's say there's a huge biome (*maybe bigger than the Pacific Ocean*), rich with plant life and fauna, including herds of dinosaurs and any animals imaginable. Suddenly, either the entire array of life is swallowed up into the ground at once, or they are all covered by a swift moving mud-wave a hundred yards thick and hundreds of miles wide. If they went straight down, then there they are encased. If taken by a moving mudbank, they are still encased, but already in motion. The mud they are in is added-to with more mudflows and more lava flows. Either way they entered the mudflow, they are all in it now: plants with human and animal remains all getting cooked by the heat of the lava that permeates and mixes with the mud. Entire bones are grabbed at once, to be mixed with other biomes.

"Into coal, oil, and gasses they go; fossils, too, don't cha know?"

Those mud flows are the new continents that rose because baked mud is loftier than unbaked mud; because lava came gushing up from the depths of the earth's loins and cooled into more mineral stuff, also expanded from its original state.

> After another forty days, Noah opened the window he had made in the boat and released a raven. The bird flew back and forth until the Floodwaters on the earth had dried up. He also released a dove to see if the water had receded and it could find dry ground. But the dove could find no place to land because the water still covered the ground. So it returned to the boat, and Noah held out his hand and drew the dove back inside. After waiting another seven days, Noah released the dove again. This time the dove returned to him in the evening with a fresh olive leaf in its beak. Then Noah knew that the Floodwaters were almost gone. He waited another seven days and then released the dove again. This time it did not come back.
>
> Noah was now 601 years old. On the first day of the new year, ten and a half months after the Flood began, the Floodwaters had almost dried up from the earth. Noah lifted back the covering of the boat and saw that the surface of the ground was drying. Two more months went by, and at last the earth was dry!
> (Genesis 8:6-14, NLT)

This might have been the most difficult time for them, as far as work goes. By now (*I'm sure*), the animals are all awake and getting a bit stir-crazy, wanting to get out and run and eat and be free. Before coming to rest on the mountain range, they were probably just hanging on for dear life in the bobbing and pitching and rolling of the ship. Now it's a totally different vibe onboard; the boat isn't moving like crazy but the animals are.

3 | The Flood

Outside, the Earth is still reeling and drifting and being reshaped forever. The Grand Canyon is forming from land being jostled that has just formed and is still fragile—not fully baked and hardened yet. Huge sheets of earth are jostling each other still across the globe, and will for a time to come; whether under the sea or rising up above it. Earthquakes are nearly constant for the Ark Crew and passengers. Mountains rise and canyons fall. We're just waiting for the world to quit reverberating from the explosion that dislodged and ruined it. Earthquakes and storms are constant reminders of the Earth's throws of agony that linger, as she is being reshaped into God's new will for her. These tremors will never leave her, for as long as she remains. She is dying now; her end has begun. Her days are numbered. Tectonic Plates are just broken pieces of the rock ball that was our original Crust Bedrock Layer. They are lying upon a ruptured and slippery Mantle that also is in motion now that it is nearly inside out. Lava separates these rock surfaces making a layer of slippery goo, we call the Asthenosphere.

> Then God said to Noah, "Leave the boat, all of you—you and your wife, and your sons and their wives. Release all the animals—the birds, the livestock, and the small animals that scurry along the ground—so they can be fruitful and multiply throughout the earth." (Genesis 8:15-17, NLT)

Grasslands

This isn't part of the model, and I don't know that it's revelation, either. I haven't read this in the Bible, and I don't think I've seen it in textbooks, personally, but it just seems to me that the world immediately following the flood would have been quickly covered in grasslands.

Grasslands everywhere

When I look at Google Earth and imagine the people's migrations from Babel to the farthest reaches of the planet, it just seems like grass would have been where deserts or barren places are now. They would have had domestic animals migrating with them. And those animals would have eaten grass. When they settled in Babel, it was called a plain. That implies lots of grass to me. They made bricks, which would require grass. Grass tends to pop up more quickly than other plants, except maybe for weeds.

I don't know, it's just a thing in my head. I see people migrating over huge grasslands on their way to their new homes (*after Babel*). And why people settled in the brown areas they are in now only makes sense if there were green grass there when they settled. Places like Babel, which now is a wasteland. That was their choice? (*Really?*) It had to be better back then compared to now. There was probably more water for crops, too. It may have been on a bay at first, as well.

My guess is that the water table has gone down over the millennia and centuries. It's been draining into the oceans, but at what rate? Did the rate slow way down at the end there and just go down so slowly that it wasn't noticed to be still receding? I don't know.

Post-Flood Weather

Okay, a quick weather review:

- **Pre-Flood weather**| perfect all the time
- **Mid-Flood weather**| more extreme than anyone but Noah and his family has ever lived through. Probably not survivable all on its own (*without shelter*).

As bad as Mid-Flood weather was, the Post-Flood weather was probably no picnic either, especially from our intrepid survivors' point of view (*having experienced nothing but perfect weather for hundreds of years—personally, in Noah's case. Shem was 98 after the flood, and Japheth was 101, while Ham was called the youngest but his birth year was not given*).

But they stepped out onto another planet completely, weather-wise. I doubt if they had any idea what they were in for, and soon. A whole new world of temperatures was upon them, forever.

Ice Age?

Really, there isn't a whole lot in the Bible about what happened right away, weather-wise. So far, we just have the mighty wind that helped with the massive runoff worldwide.

(*The runoff that continues to this day, in my opinion. Since my earliest memories, I've seen the water table dwindle across my part of the planet. We are slowly losing our good, potable water. Minerals and other water-soluble elements are leaching into the water below ground, making it largely undrinkable.*)

Weather-wise, I think that's about all we get from the Bible in early Earth history. But we know from history and the geological evidence that there was a period of heavy, heavy snowfall not long after the Flood was over (*probably during or*

3 | The Flood

immediately after the crazy weather of the Flood). Actually, I think that it was **during** the Flood that the snow and ice began to fall on the poles. I think that it snowed to perhaps the 40th parallels, north and south. This is due to the atmosphere having more water content then than at any other time in history. That would make the accumulation inimitable (*unrepeatable*).

I'm not a meteorologist but I was trained a little in weather. What I think happened was the cold trend continued to the point that snow and hail and ice began to fall so heavily, throughout so much of the world, that it froze much of the planet for at least a few years if not decades, as the seas cooled and weather patterns were established and stabilized. Weather that severe would not be possible at any other time in history, before or after. Looking at Antarctica and the Arctic Circle, Earth's two largest deserts, it would be impossible for them to have any snow or ice at all except for this meteorologically unique time. There had to be the combination of sustained cloud cover with unmatched precipitation (rain, snow, hail, and ice).

Fortunately, our intrepid survivors landed in an area where the snow and temperatures weren't that bad; they didn't land on the mountains of Antarctica, for example. And they didn't land someplace impossible to navigate from; like the middle of the Himalayas, or the Congo jungle. From the mountains of Ararat, they could have easily gotten to decent latitudes where the temps were livable. (*Saying they were on the mountains doesn't mean the highest point.*) Personally, I think that they were on the southern slope of Ararat about a thousand feet from the top. Look on Google Earth for what looks like a huge footprint on that part of the big mountain. They would have landed just uphill from that.

That's when the polar regions, North and South, froze over. It's no secret that there were many years of what we call an "Ice Age". What we can't seem to agree on is how many ice ages we've had and how long they've lasted in all. Personally, I don't think that there has been any more than one "ice age", but that it has just lasted for a while in the northernmost or southernmost reaches of the planet (*I guess it's still going on at the poles, isn't it?*) As covered as Antarctica is with snow and ice, it's actually the largest desert on Earth. This means that there is no way that it could ever have enough snowfall to repeat its conditions of today. In fact, it will only continue to melt away. But it will not add to sea level until a water table is first established under the surface of the southern continent.

Did the glaciers oscillate north and south a bit as they retreated? Maybe. But they retreated steadily after they were first formed, not long after the Flood, and still going. Oh, and don't confuse "glacial scars" with flood scars. Rocks show signs of rocks being dragged across them, leaving huge scars. Some scarring could occur from ice moving as well, I guess, but the flood did much more damage to the surface than the retreating, melting ice sheets.

3 | The Flood

I want to add, really quickly, that the Lord will probably be back before Greenland is green. Just keepin' it real.

Tower of Babel

We will look at their migrations in another chapter, but from here Noah's family moved South (*to escape the cold?*) and a bit East and eventually began to build up a city in ancient Mesopotamia, which is modern day Iraq, when God reminded them that they weren't doing what He told them to do—namely, spreading out!

That was the Tower of Babel Incident that happened to Noah and his family, all the way down to great, great, great grandson, Peleg, who was born in 1,758, AM (*anno mundi, year of Earth*). His name means "division" because he was born during the great division of people and language. (*See Genesis 10:25.*)

If 1,758 is when the division happened, then that was a mere 101 years after the flood was over in 1,657. So, 100 years after the flood, everyone had their lives turned upside down all over again. Was God being unfair or mean? I don't think so. If He hadn't broken everyone up, they would've turned society into one big city that would've just kept growing and growing. I don't see that as being the best way to grow the population (*in a crowded urban environment*). Plus, it's cool how all civilizations have something unique about them. And I like the variety of languages and ethnicities.

By the way, Noah outlived Peleg, his great, great, great, grandson, by about 9 years. Peleg died in 1,997 *am* and Noah died in 2,006 *am* (*3 years before Abraham was born in 2,009 and 352 years after the flood*). Peleg was 239 years old; Noah was 950. "Radiation, radiation, radiation." Soon after that, God's edict of no one living more than 120 years was realized.

Yes, Noah and his sons were presumably all there in Babel during the great breakup of their family into clans and societies. They probably lit a fire under everyone to go as far and as fast as possible when the languages hit. And they did—probably ill-prepared and ill-equipped. You can imagine why Noah didn't want to test God's patience another second. He had already seen what happens when people ignore Him. You don't need language to communicate to your family, "Go"! "Go far"! Noah was 702 and probably not going anywhere, now that radiation was zapping his energy. And if a Nimrod-type brute were pushing people out, eesh! It could've been brutal. I think that Nimrod came later, but who knows? It could have been a very sweet, tearful, and loving goodbye.

3 | The Flood

This is when various **religions** began to pop up all around the world (*because people don't seem to like repeating sad stories—like a good God wiping out almost all of humanity in a flood because of their evil ways, I guess*). God's methods are so very clever and effective, aren't they? I can see it now: God ~ "These guys aren't breaking up like I said. I know, I'll make it so they can't talk to each other, that'll split them up." Brilliant! It worked perfectly. They all split up right away and some of them went many thousands of miles before putting down roots that would establish cradles of humanity halfway around the globe. They probably all went away angry (*at God*), sad, or scared about what happened and what might happen next.

Ethnicities

This is the beginning of **ethnicities**; differences in human appearance based on genetic lines of separation. Because this caused a shallower gene pool for the various family lines, "ethnic traits" began to appear in the family everywhere. The ones who went to Africa, for example, never again saw their family who went to the Americas in the east (*because they went eastward to get there*). When they spread out, they really spread out.

Some of them went straight to what is now Central America, across arctic ice at the Bering Strait or sailing the Pacific Equatorial Counter Current, to become the Olmec, Aztec, and Mayan peoples. Some of them stayed close and populated Mesopotamia or went West and became Egyptians

127

or to the North to become the Minoans; some formed cradles of civilization in what are now China or India. On and on it went, all over the globe, farther and farther they spread out. I think that the Mayan culture was being constructed while Noah still lived in Babel. It could have been that quick.

At the birth of every language (*at Babel*), all people were of one cloth. They were one big family under Noah, the righteous patriarch. Imagine Asian language speakers looking exactly like Germanic language speakers, or African or any other. One family with much capacity for genetic variety in appearance and other qualities.

I think that the real "human race" was a footrace to see who could build the tallest ziggurat type building in the world as quickly and as far away from home as possible. After they had all agreed to build a really monumental structure back home to make them "famous" (*at Babel*), they probably kept that spirit with them as they all parted ways. The Bible says nothing of it going away from the people. Most early cradles of society have some kind of ziggurat or pyramid in their architecture, made of bricks. It was "in" when they moved out. I guess you can take the boy out of the ziggurat; but you can't take the ziggurat out of the boy. ¯_(ツ)_/¯

3 | The Flood

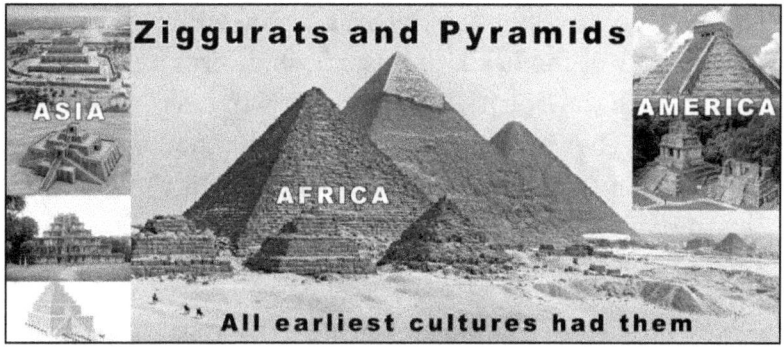

I don't believe in racism because I don't believe in races. There is one race on Earth—human. All of the differences in our appearance are due to being split up a long time ago. Racism is just ignorant hatred directed at people who look, act, smell, talk, dress, or anything else, different(ly). Acting like that toward our fellow brethren is offensive. Don't do it. It shows your lack of godliness. We are all one family.

Godliness, in case you're not sure, is being good. It's also called being "righteous". We are godly when we act like God would act. God is good.

Here is a quick teaching from Paul on this:

> You, my brothers and sisters, were called to be free. But do not use your freedom to indulge the flesh; rather, serve one another humbly in love. For the entire law is fulfilled in keeping this one command: "Love your neighbor as yourself."
>
> If you bite and devour each other, watch out or you will be destroyed by each other.

> So I say, walk by the Spirit, and you will not gratify the desires of the flesh. For the flesh desires what is contrary to the Spirit, and the Spirit what is contrary to the flesh. They are in conflict with each other, so that you are not to do whatever you want. But if you are led by the Spirit, you are not under the law.
> (Galatians 5:13-18)

May I just stop right there? You are not allowed to do just whatever you want. Don't act out on that negative thought or fantasy. Don't say those hurtful words. If you can't contain or control yourself, move to another location until you can. In another letter, Paul called this "fighting the good fight". It means contending with the desires of the flesh. Fighting our own depraved nature. Maybe if they had this teaching in Noah's day things would be different. (*"Nah, We've got it now and just look at us."*)

> The acts of the flesh are obvious: sexual immorality, impurity and debauchery; idolatry and witchcraft; hatred, discord, jealousy, fits of rage, selfish ambition, dissensions, factions and envy; drunkenness, orgies, and the like. I warn you, as I did before, that those who live like this will not inherit the kingdom of God. (Galatians 5:19-21, NLT)

Obviously, that paragraph could be turned into another book series. And I don't want to harp on various sins common to man. I don't think that Paul meant you had to do all of them to be guilty; I think he meant any one of them, as a character trait or habit or lifestyle or the like, is enough to make us

3 | The Flood

guilty (*and these are just examples, not a comprehensive list*). When we act like this we need to stop and fight the urge or desire to ever do it again. All people sin. But not all people repent (*feel sorry about it and turn from it back to God*). God tells sinners to repent. When we do, He forgives; when we don't, He doesn't forgive. And He judges us for not forgiving each other too—quite heavily.

> But the fruit of the Spirit is love, joy, peace, patience, kindness, goodness, faithfulness, gentleness and self-control. Against such things there is no law. Those who belong to Christ Jesus have crucified the flesh with its passions and desires. Since we live by the Spirit, let us keep in step with the Spirit. Let us not become conceited, provoking and envying each other. (*Galatians 5:22-26, NLT*)

Sadly, some of the groups may have abandoned God right away. Maybe they were disenchanted from losing so many of their family and friends and way of life, idk. All they knew was Noah and his family. And then he sent them away, probably in haste; great haste—too hastily to get ready for the trip. You know, taking instructions from the others who know stuff they don't. Getting all the stuff they'll need for the journey...nope, just "GO! QUICKLY! BEFORE GOD KILLS US ALL FOR DISOBEYING HIM AGAIN!—GO!!" (*Seven-hundred-year-old Noah, shuffle-running through the city; both hands waving over his head; screaming to his progeny as he goes... Just guessing.*) They didn't even know how small the new world was at the time compared to before the flood. Many of us today don't know.

Cavemen

So far, we've seen that a family of eight survived the worst catastrophe to ever hit the world. In fact, they were the first humans to travel in a ship from one world to another.

This new world they landed on was a lot more hostile, geologically and meteorologically, than their previous world that was now gone forever. Ruined. Destroyed. Lost.

All the technology, knowledge, and skill of humanity was in their 16 hands—they brought it with them; in their heads and hands (*and writings?*). The rest was lost. I'm sure that they took a hit in their technological progress from all of this. But then they started to rebuild their society. And just when they were getting more advanced in what they knew how to do, while building the tower of Babel, BAM!—it all has to go away—again. They all had to go back two spaces on God's game-board of life.

3 | The Flood

This sent them in all directions with little more than the clothes on their backs and what they could carry, cart, drag, or load onto animals or boats. No doubt, some of them knew more about life than others did. Some of them could do things that others could not—like metallurgy, or textiles, or pottery, building carts, etc.

Is it any wonder then that we find evidence of migrant people living in caves or other natural shelters in a very simple lifestyle without a lot of technology? Is it any wonder that all early societies had rituals for burial or marriage or birth?

It's no secret that those who found themselves in the Neander Valley *("Neandertal" in German)*, for example, had a different appearance than most other people. They used rocks as tools because they didn't know what others knew about metals. Eventually, their offspring will figure it out or learn from others they meet. (*And their automobiles will be very popular someday.*)

Now, isn't this story very different from what you heard in school about this one branch of our family tree? We were told that the Neanderthals were primitive cavemen; a link between humans and apes; barely able to speak, if at all. They just grunted and motioned with each other to communicate; not really able to form sentences.

Well, that was a lie. The Neanderthals were a branch of the family tree who were just like the rest of us, only with more prominent eyebrows than the others, perhaps. Go back to the *Apes Following Men* cartoon, Frame 1, and look at the guy who's speaking the dialogue. He's a Neanderthal, I think. Whatever he's supposed to be, he looks just like the other two—manly. And he made huts to live in with his family. And he traded with his larger family—the rest of humanity (*nearby*). I mean, the ones who knew metallurgy went one way and the ones who didn't went another (*probably wishing that they had a blacksmith in the family*). The same is true for the other skills: carpentry; pottery; weaving; so on and so forth. We don't see these skills popping up in various places; we see these skills being taken to various places from Babel or developing further afterward.

We know they had religion and art and higher ways than animals, which all people exhibit. They weren't primitive biologically; they just lacked some of the technology that they once used to enjoy before the division. They were of the same character as us.

I guarantee, if you had to survive today being taken from your society while traveling all over This Broken Planet, you would be lacking a few of the more stable conveniences of life. You'd live a "primitive" lifestyle for a time. Your situation might even be called "apocalyptic".

As a person of Germanic descent and prominent eyebrows, I am likely a member of the Neanderthal family tree. Maybe that explains a lot about me, I don't know. ¯_(ツ)_/¯

3 | The Flood

Great Flood Synopsis

1. Earth's Mantle breaks somehow causing lava to come out, thereby exposing everything above to:

a. Extreme heat

b. Radiation

c. Spreading of hot, radioactive lava worldwide under, in, and on the Crust Layer

2. The Mantle's breaking releases lava that contacts the water and causes the Crust to break due to an explosion of water vapor.

3. The Crust's breaking allows the Water Layer to be displaced, sending the Crust's bedrock to the Mantle in a crumbled mess. Pieces can be gigantic or small, stacked or layered, or overlapped; upside down or right-side up, lying on the Mantle or falling into it, riding on magma and melting into it. Subduction is introduced. Tectonic plates are created. The Asthenosphere is created and starts to grow. That's the molten, squishy, liquid layer of really hot stuff just under the tectonic plates. The more we get exposed to that, the hotter it will get, I'd think. (*Global melting?*)

4. At the same time, over the course of about 5 months, the soft soil is redistributed into the continents of today by water dissolving and liquefying it, and the heat (*from the magma and lava*) cooking it into rock, in-situ. This means

that the land fell out from under everything and was replaced with scalding water/mud that seemingly came from nowhere. It was a quick death, globally. This is not the rainstorm that people have been saying it was. It was two parts of the Crust doing two different things: 1. Rock Falling; and 2. Soil Swirling. After that came the rain.

5. Everything that we can observe in the world geologically testifies to this story being true as just told. Mostly, EROSION just screams of the flood. You know, "Flood, Flood, Flood!" like on and on and on…it just won't shut up about it. But deep time isn't anywhere on Earth to be found…just crickets| *created critters creeping in crags, chirping cheerful chimes.*

6. After the Flood, Noah and the boys and women, went down south and east to a place in modern Iraq (*probably to escape the Ice Age*). This became Babel. (*Babel is where Noah saw what God had done with the languages and probably made everyone get out too quickly to prepare—that's conjecture on my part*). But they did spread out from Babel rather quickly and globally. He must have given them instructions to go as far as possible from him. "Go, until you can go no more!"…"Don't stop for twelve moons!" or whatever. I mean, being family they still had ways to communicate certain simple things.

3 | The Flood

Group Discussion Questions:
1. Was the Flood primarily water coming up from below or from rain in the sky?
2. Why was the one Ice Age created and why can it not be duplicated?
3. If all of the snow and ice on the polar caps melted today how much would the oceans rise?
4. What caused various civilizations and ethnicities around the world?

Answers:
1. Water coming from under the ground.
2. Because we will never again have such cloud cover and that much precipitation again. Especially since the polar regions (Arctic Circle and Antarctica) are the two largest deserts on Earth.
3. Zero, since the ice in water does not make the water level rise and the snow and ice on ground would first create a water table.
4. The division and migration of Noah's family due to the Tower of Babel incident.

Chapter 4 | "Earth Looks Old"

"Why does Earth look old?"

"Because it's broken and falling apart".

Even the newest toy or tool can quickly become old looking when it is broken beyond repair, especially if it were of particularly fragile design to begin with. And the Earth was of very fragile design when new and treated very harshly on its entire surface.

Earth didn't have radiation destroying the life and minerals before the Flood. But now that its inner radiation has been released, everything looks bad—because it is bad—and getting worse! This is because radiation destroys any life it contacts and even rocks decay from it.

We call this "Entropy"—the tendency of everything to breakdown. Things are getting worse; not better. Things are getting simpler; not more complex. Things are breaking down and decaying; not growing and becoming better and stronger. (*Funny how this law of physics is ignored by those who say things got more complex and better over time—huge amounts of time that were never there.*)

This Broken Planet

Where I live in the southwest US, jagged Rocky Mountains loom over the city. Those weren't there in the days before Noah's Flood, discussed in Genesis of the Bible. They appeared during the Flood when huge masses of new earth were shoved into each other and wrinkled up from east and west to form the mountain range we call "The Rockies". (*Not to mention the Cascades and the Sierra Nevadas.*) But none of the American mountain ranges existed before the Flood. That would include all of these: North America
1. Brooks Range (Alaska, USA)
2. Alaska Range (Alaska, USA)
3. Saint Elias Mountains (Alaska, USA; Yukon, Canada)
4. Coast Mountains (British Columbia, Canada; Alaska, USA)
5. Cascade Range (British Columbia, Canada; Washington, Oregon, California, USA)
6. Sierra Nevada (California, Nevada, USA)
7. Klamath Mountains (Oregon, California, USA)
8. Peninsular Ranges (California, USA; Baja California, Mexico)
9. Sierra Madre Occidental (Mexico)
10. Trans-Mexican Volcanic Belt (Mexico)

Central America
11. Sierra Madre de Chiapas (Mexico, Guatemala, El Salvador)
12. Cordillera Isabelia (Honduras, Nicaragua)
13. Cordillera de Talamanca (Costa Rica, Panama)

South America
14. Andes Mountains (Venezuela, Colombia, Ecuador, Peru,

4 | "Earth Looks Old"

Bolivia, Chile, Argentina)
• Cordillera Occidental (Western Andes)
• Cordillera Central (Central Andes)
• Cordillera Oriental (Eastern Andes)
15. Cordillera de la Costa (Venezuela, Colombia, Ecuador, Peru, Chile)
16. Cordillera Blanca (Peru)
17. Cordillera Negra (Peru)
18. Cordillera Darwin (Tierra del Fuego, Chile)

As beautiful as they are, they are wrinkled-up piles, turned to rock, of what used to be the perfect soil of the antediluvian world (*the one before the Flood*).

The Flood is why Earth looks old. It damaged Earth to the point of no return to its early glory and beauty. (*And yes, it is still beautiful in its dying, decaying state.*) It still nurtures life, but just barely; a mere shadow of its early, vibrant splendor.

A Poetic Interlude (and ode to the Lord)

Let all that I am praise the Lord.

O Lord my God, how great you are!
You are robed with honor and majesty.
You are dressed in a robe of light.
You stretch out the starry curtain of the heavens;
you lay out the rafters of your home in the rain clouds.
You make the clouds your chariot;
 you ride upon the wings of the wind.
The winds are your messengers;
 flames of fire are your servants.

This Broken Planet

> You placed the world [Crust?]
> on its foundation [Mantle?] so it would never be moved.
> You clothed the earth with floods of water,
> water that covered even the mountains.
> At your command, the water fled;
> at the sound of your thunder, it hurried away.
> Mountains rose and valleys sank
> to the levels you decreed.
> Then you set a firm boundary for the seas,
> so they would never again cover the earth.
>
> You make springs pour water into the ravines,
>   so streams gush down from the mountains.
> They provide water for all the animals,
>   and the wild donkeys quench their thirst.
> The birds nest beside the streams
>   and sing among the branches of the trees.
> You send rain on the mountains from your heavenly home,
>   and you fill the earth with the fruit of your labor.
> You cause grass to grow for the livestock
>   and plants for people to use.
> You allow them to produce food from the earth—
> wine to make them glad,
> olive oil to soothe their skin,
>   and bread to give them strength.
> The trees of the Lord are well cared for—
>   the cedars of Lebanon that he planted.
> There the birds make their nests,
>   and the storks make their homes in the cypresses.

4 | "Earth Looks Old"

> High in the mountains live the wild goats,
>> and the rocks form a refuge for the hyraxes.
>
>> You made the moon to mark the seasons,
>> and the sun knows when to set.
> You send the darkness, and it becomes night,
>> when all the forest animals prowl about.
> Then the young lions roar for their prey,
>> stalking the food provided by God.
> At dawn they slink back
>> into their dens to rest.
> Then people go off to their work,
>> where they labor until evening.
>
>> O Lord, what a variety of things you have made!
>> In wisdom you have made them all.
>> The earth is full of your creatures.
> Here is the ocean, vast and wide,
>> teeming with life of every kind,
>> both large and small.
> See the ships sailing along,
>> and Leviathan, which you made to play in the sea.
>
>> They all depend on you
>> to give them food as they need it.
> When you supply it, they gather it.
>> You open your hand to feed them,
>> and they are richly satisfied.
> But if you turn away from them, they panic.
>> When you take away their breath,
>> they die and turn again to dust.

When you give them your breath [Spirit], life is created,
  and you renew the face of the earth.

  May the glory of the Lord continue forever!
  The Lord takes pleasure in all he has made!
The earth trembles at his glance;
  the mountains smoke at his touch.

  I will sing to the Lord as long as I live.
  I will praise my God to my last breath!
May all my thoughts be pleasing to him,
  for I rejoice in the Lord.
Let all sinners vanish from the face of the earth;
  let the wicked disappear forever.

  Let all that I am praise the Lord.

  Praise the Lord!

  (Psalm 104)

Conflicting Cosmologies

For the last couple of hundred years since the Enlightenment of the 18th century, people have been trying to put evolution into biblical cosmology and the Bible into evolutionistic cosmology.

How has this worked out so far? It hasn't. There aren't any resolutions to be found between the competing worldviews of secular, evolutionistic cosmology (*which depends upon deep time*) and biblical, young earth creationistic cosmology (*which demonstrates a young earth*).

4 | "Earth Looks Old"

So, whichever side you've been on lately, you should admit that it hasn't been working if trying to bring them together. Maybe I can help you with that. First, stop trying. If you think it's working, you should know that it's a delusion. An impossibility. They are diametrically incompatible. It's a target that will never be hit. You're wasting your time. Give it up.

One thing that must be dropped from cosmology is this concept of "Deep Time". That's what we call the story you've been hearing your whole life, that "the world is something like 4.5 billion years old". This idea of Deep Time was popularized in Edinburgh, Scotland in the 1,780's.

James Hutton

James Hutton, called "the father of Geology" by his pals in Edinburgh, penned some thoughts in 1,785 that some folks of his day were eager to accept. He basically took God (*and His Church*) out of the picture by imposing strict guidelines into everyone's thinking on Geology. Here's how he put it:

"The past history of our globe must be explained by what can be seen to be happening now. No powers are to be employed that are not natural to the globe, no action to be admitted except those of which we know the principle."

145

> (~ Hutton's Theory of the Earth was presented in 1,785 in front of the Royal Society of Edinburgh, then published in 1,788 and enlarged to two volumes in 1,795.)

And then he went on to add:

> "The result, therefore, of our present enquiry is, that we find no vestige of a beginning—no prospect of an end." (Same source)

Whatever he was thinking and believing when he wrote this is not known to me. I never met him or his buddies nor read anything more from or about them, really. I'm no expert on the Enlightenment either. But I can see what he was doing in making such statements. He was stacking the deck; making it impossible for the Creation story to be considered as equal to his and his buddies' ideas. Charles Lyell picked up on it too. I think Hutton wanted to begin the debate with a claim of **eternal existence** for Earth, so that bringing it down to a ridiculously huge and inconceivable number may be easier to swallow. A compromise. He *was* a lawyer, after all; trained in hostile negotiations, no doubt.

If Hutton and the other influencers of the enlightenment were attempting to get the Church out of their social club, then they did a pretty good job of it over time. Today the Science Club and the Bible Club hardly even talk to each other, much less share their hopes and dreams and doctrines with each other.

4 | "Earth Looks Old"

As it turns out, though, Hutton was completely wrong about the past NOT being the key to understanding the present (*and future*). He wanted to explain the past based on the present. But that doesn't work when the past had a catastrophic event that reshaped and ruined the planet irreparably. He and his social club buddies talked the world into seeing things completely backwards, leading to the most inaccurate results that could have been realized. People like Chuck Lyell and Chucky Darwin only added to it in the years following. That bandwagon has a lot of disciples, sadly.

Because the world fell for this story and way of approaching science (*and God*), we have been chasing a lie for over 200 years! We are so far from the truth now that those who have been raised in this doctrine of ignorance can only laugh at the prospect of any other paradigm. But again, it's not their fault that they've been spoon-fed poo their whole lives and accept it as good and true. (*Not judging the peeps.*) But it would be nice to get this bag of ignorance off the heads of well-meaning scientists.

I see people following Hutton's way as tantamount to putting a bag over their head before engaging in science.

What's wrong with "Deep Time"?

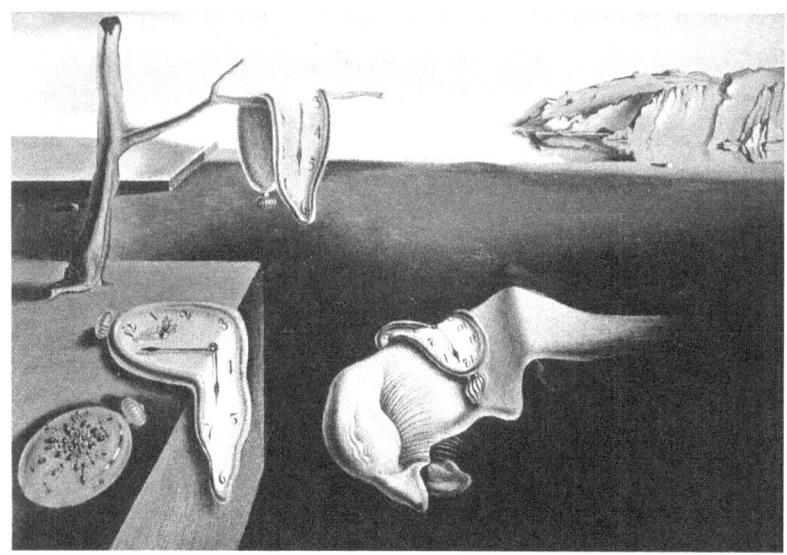

Salvador Dali—*The persistence of memory, 1931*

A lot is wrong with deep time. It is nothing more than a sham; a lie; and false pretense that cannot work in the real world. There is absolutely ZERO evidence of long periods of time on Earth. In fact, there is the exact opposite when tested scientifically and honestly. Please, let me explain.

When we observe layers of rock, sand, clay, ash, and whatever else is stacked up in the "soil" before us (*like looking at the walls of the Grand Canyon, for example*), we can see that there are huge deposits of these minerals and more that have been apparently laid down by the force of moving water (*mud, really*).

4 | "Earth Looks Old"

Where any two layers are stacked up, the Deep Time explanation has been that the layers were put down over millions or billions of years, very s l o w l y. Why do they say this? I don't know. It doesn't make any sense to me, whatsoever.

Here are a few scientifically observable reasons why there could NOT have been long periods of time between layers of soil or rock deposition, and why evolutionists and "Olde Earthers" are going to go sit down on the Flat-Earthers' bench when we hold them accountable (*this is science; not Bible*; *it absolutely guts both concepts*):

Disturbance

I'm going to make an outrageous statement that I will clarify as we go:

Any stratum that exists under other strata does not show any signs of disturbance, anywhere in the world.

I know that this is quite outrageous to some, but it will make sense before too long just what I mean. Just stay with me for now. No burrows of animals or insects. No erosion. Nothing. How can this be after m/billions of years and all over Earth? It looks as though the top layer came so quickly after the first that nothing had time to disturb it—not a rodent, a reptile, a bird, or a bug (*or a dude*). Actually, they were all dead at the time, sadly.

Burrows, tracks, and erosion are mostly absent where two or more strata have been stacked up. (*In some places, a few footprints are in "older" rock than their makers' fossilized feet in "newer" rock—oh boy.*) Their timeline is make-believe. All we see in the mud-rock layers is petrified dead things; not their homes. No other evidence of them in the soil, like we see today.

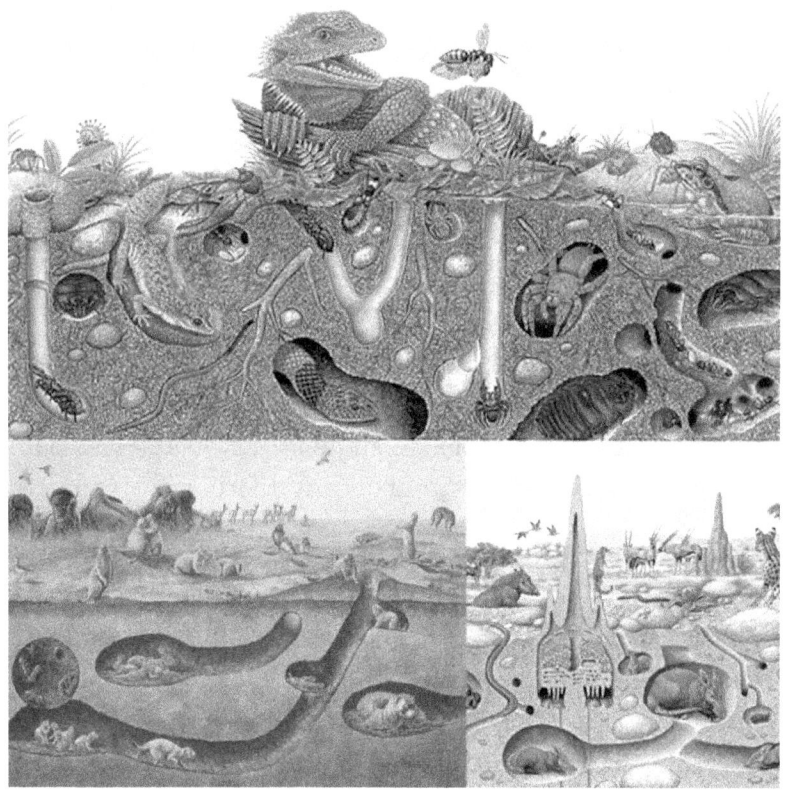

Compare top images with bottom pic; see any disturbance between strata (*fine lines*) below? Me neither.

4 | "Earth Looks Old"

Remember, the lie has been that each little stratum above was formed over millions of years. Eesh.

Polystrate Fossils

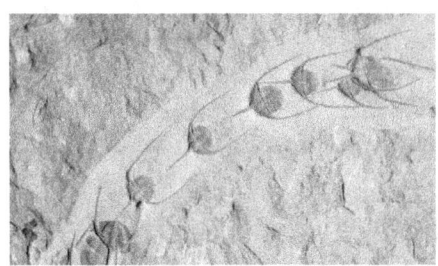

When we look into the rocks under our feet, we see that Trilobites have died and been fossilized while crawling through the mud. They only made it a little way before succumbing to the mud and heat, but they were preserved in their struggle to get uncovered by the flow of hot mud that encased them quickly.

"Pretty amazing thing to occur over a million years!"

"Uh, no, it didn't happen that slowly."

Yeah, they were covered in *seconds to minutes*, not weeks, months, years, decades, or centuries, and definitely not millions of years. They were covered up in a second in mud and then remained encased forever. That's how fossils are made—dead things in hot mud turning to rock. It's a baking process. A quick one, at that.

Vertical fossils of trees run through multiple "layers" (*I'm using the word as loosely as the textbooks here*). There are several examples of trees and other forms of life or rock passing right through several layers. And some layers are all bent in a swirl together.

"How did that happen over millions of years?" Answer: "It didn't".

It happened quickly in forming (*baking*) rock (*while still soft and not yet hardened—think of gooey bread, still baking*). But those fossilized trees did not grow where they're encased. They were deposited, along with everything else in the mud. The roots are largely gone; broken off in the flood.

4 | "Earth Looks Old"

Wavy Strata—formed before fully hardened.

Polystrate Fossil—punching through many strata.

This phenomenon is not helping the deep time position any. In fact, it cuts right through it, all around the world. Just like the tree fossils cut through their "millions of years" worth of layers, this fact cuts right through their story that was just a figment of someone's imagination (*James Hutton*) and repeated for over 200 years. Sounds kind of like a myth or a fairytale in the making, doesn't it? But since people are making a living off of this and it's deliberate deception, I'm calling it a $cam. Especially since it's been forced upon us, and at odds with the truth.

Those layers all around the world hardly have *days* between them, much less *years*. Anything beyond that is silly. The bigger the number gets; the sillier it is. One million years between layers? Very funny. Ha, ha. Ludicrous.

*(**Side note:** Layers and strata are confusing terms that seem to be interchangeable in the schools. If it were mine to say, I'd have "strata" be the little lines that appear in dirt or rock that has been immersed and moved by water. "Layers" would then be the different flows of mud that were stacked up all over the world. So, in my mind that's what they are and I thought you should know that, since you're reading my stuff here.)*

The actual layers, in my opinion, are huge mega-flows of mud that we'll look at in a little bit—Mud Tsunamis cooked to rock.

If you want to see what "strata" look like, take some dirt and put it in a jar (*just one third to halfway full*). Fill the jar

4 | "Earth Looks Old"

with water and make sure that all the dirt gets diluted with it. Shake it up and let it settle. In a short time (*hours*), you'll see "layers" form, just like the ones in the Grand Canyon.

Jar of water test

If you did that, you did a science experiment that shows how **strata** are formed naturally in dirt when carried in water. The mud strata remain when hardened to rock by heat.

"Gee, that's just like what happened in the Great Flood."

"Millions of years between strata? Give me a break!"

155

We know better now. Mt. St. Helens (*1980*) taught us that these things happen very rapidly. (*Thank you, Dr. Steve Austin, PhD.*)

Stratification is just a natural phenomenon associated with flows of minerals in water. And not just trees, but the many bones of segmented animal remains or the complete bodies of some animals are carried in the same manner. All kinds of animal remains, trees, and other flora have been encased in rock as well—as fossils or fossil fuels. Many of them cut right through the rock strata that are said to be laid down very slowly, over millions of years.

Everyone agrees that water covered every inch of the globe at one time. All of our sedimentary rock shows signs of stratification from water flow. This is universal around the globe and the timing matches everywhere as well.

Fossil Fuels Recipe:
Dump biomass into a boiling glob of mud.
Cook it under pressure for a while (weeks to years).
You'll either get coal or oil or fossils, depending.
You may also get a bunch of gas.
"*It's not that complicated. Just don't ask a professor to explain it, unless she's a biblical creation geologist.*" (ツ)

4 | "Earth Looks Old"

The men are standing on one layer looking at another. Dr. Steve Austin (*hand raised*) is a prominent Creation Geologist Professor with a PhD in Geology. He was instrumental in teaching stratification at the Mt. St. Helens eruption site. Image taken from the film, *Is Genesis History?* which is available for free on YouTube. I highly recommend it. They didn't have the revelation given to me but they can read the signs in the earth and know that this happened quickly.

Erosion

Erosion is so severe and prevalent worldwide (*mostly on the surface*) that nothing but a worldwide flood can explain it. The same is true of the Continental Shelf that exists around every continent.

"Wait! Didn't you say that there was no erosion?"
"I did."
"Aha!"
"Between '*strata*'. *Look at your dirt in the jar.*" But yes, there

157

is a huge amount of erosion, but where it is and how it appears is what is important. Stay with me.

These flows all show signs of occurring in the same time period—just like a worldwide flood would do. They all occurred in a few months' or weeks' time of each other, tops. More likely it was hours or days between flows, lasting for many months. The erosion on Earth is consistent across the surface. It lasted over 150 days straight and years following.

And when we see how some of the surficial rock features formed from erosion are precariously balanced, or are of a particularly crumbly, fragile nature, we can deduce that those features formed quickly in the recent past. Otherwise, it isn't likely for some of the precarious and delicate formations to still be perched where they are. If you don't know what I mean, look at the Badlands of the Dakotas in the US. (*Or look at the picture.*)

4 | "Earth Looks Old"

Fossils and Fossil Fuels

Fossils and fossil fuels, like Coal, Oil, Gases (*Methane and Natural Gas*), etcetera, could have been formed very quickly in the immense pressure and heat that was holding them. Just like the fossils, the biomass that is now coal or oil or gas was wrapped in a mud-bake to be pressure-cooked.

This mud-bake was very quickly setup by the coming together of a few important, inimitable (*unrepeatable*) conditions:

1. In a worldwide earthquake (*explosion*), the Crust bedrock fractures and falls into the Water Layer below, coming to rest on what's left of the searing hot Mantle at the bottom, which has had its surface blasted into its innards. The Mantle was blown inward in places, while the Crust was blown outward before collapsing back onto the imploded Mantle. The Crust pieces are now exposed to lava and magma from the ruptured Mantle, which is melting the soil and some of the rocks it contacts, making more lava. The Crust and atmosphere are irradiated by the escape of radiation from below. The radiation is destroying the broken rocks too. The lava is melting them in places. That's global melting; the rising of the Asthenosphere! It might not stop. But that's okay. Jesus will be back before

Greenland is green. Seriously, who knows how much lava will continue to ooze forth until it stops?

2. The soft soil that was on top of the bedrock is quickly incorporated into the water that is suddenly escaping the fractured, dissolving Crust Layer. The Water Layer was in motion before this explosion. That means that the situation is much more dynamic than one might imagine. The heat of the water causes the soil to dissolve quickly into mud as the flows churn it violently. This mud is what all the people, animals, buildings, plants, sea life, and everything else were living on just seconds before being swallowed up in this burning mud that is churning and turning, bubbling and boiling. Prior to the flood, the flow of the water went unnoticed miles below. And into it they all went—all life! In less than an hour, I suppose, all life on earth was being sloshed, tossed, broken up, and cooked in the mud by the heat under it all. It was a quick death for everything. I'm amazed that any sea life lived at all. The Ark story is a miracle! Such turbulence!

3. This mud, that was solidifying from the massive heat under and around it, was quickly shaping into built-up heaps that will together form the continents. Layer after layer of thick mud, carrying the remains of Earth's lifeforms, began to solidify into these massive layers; eventually becoming dry land covering a mere 29% of the surface (*where the soil used to be nearly everywhere on the surface, minus some lakes and seas*). This is how the continents were formed in the water with lava and mud

4 | "Earth Looks Old"

expanding as they cool, harden, and meld. The former soil layer becomes the new sedimentary rock and the lava becomes igneous rock.

4. The hot Water Layer is no longer protecting the soil from the immense heat of the Mantle because it was escaping due to the falling Bedrock pieces. This heat is now doing what any kiln or potter's oven would do; it's baking the mud in place. And as more mudflows are caught and added on, the fledgling continents grow in height (*mud expands when it bakes, I would imagine*). I think God was ensuring that we would have enough land protruding above the waterline at the end of it all. So the soil was scooped into piles as it baked. I suspect that it takes both lava and mud to make a continent. The lava (*and magma*) is coming up from the depths of the Mantle in certain spots, and the mud is what's left of the original soil. Together, they are what make up the continents. The Mid-Atlantic Ridge is where the lava is still oozing out from between two plates touching. Many islands in the Pacific, for example, seem to just be cones of lava that shot straight up to the surface. Lava is a big part of the recipe.

And so, with the Water Layer migrating to above the Crust Layer, and the continents being built up from the mud and lava turning to rock, everything was put into the configuration we see today all over This Broken Planet.

Mud Flows Into Rock

In just a few years, really, all of the coal seams and oil fields on Earth could have been made from the intensely populated biomes of life that were covered, squeezed, and wrapped into the mud, then cooked during and after the massive tsunamis of the great deluge, called Noah's Flood. It all happened very quickly and violently. Nothing had a chance to survive. Just some sea life and those in the Ark with Noah.

There was so much life on Earth before the Flood that if another were to happen today, we wouldn't get anywhere near the same amount of these fossil fuels. In fact, I've heard that there is not enough life on the planet today to even come close. But of course we can't get the conditions we have today from the conditions we have today, now can we?

For example, look at this image below. What I'm saying is that everything between the orange start line and the white finish line was some kind of forest or other biome just packed and teeming with life. The scrape marks on the ocean floor seem to indicate that much geo-mass was scooted along the floor to where it's obviously piled up into the Sierra Nevada, Cascades, and Rocky Mountains. And if it did all that, then making the Grand Canyon would be a sneeze. And that's just the part that gave the US its fossil fuels. Imagine the vast jungles and forests and swamps that used to lie between Denver and Honolulu, before the Pacific Ocean was in place. Can you see all the scars of volcanism in the Southwest USA (*Utah to NM and AZ*)? This is physical evidence of what I'm saying here. It was huge amounts of lava and mud being pushed along by extruding water from under the huge plates that fell after being fractured.

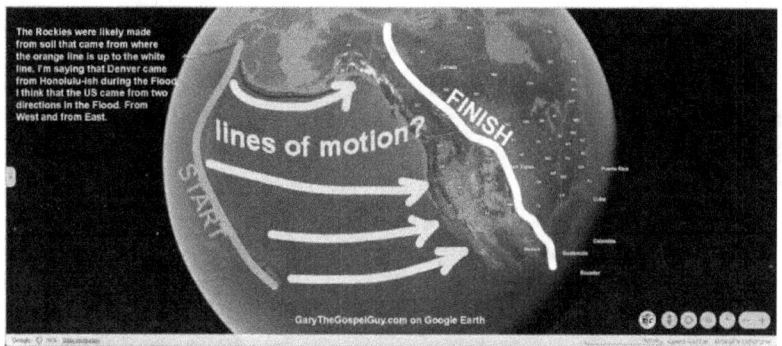

CAPTION: *The Rockies were likely made from soil that came from where the start line is up to the finish line. I'm saying that Denver came from Honolulu-ish during the Flood. I think that North America came from two directions in the Flood. From West and from East.*

This is what happened when the Mantle ruptured and the Crust cracked and sank: Lava came out of the Mantle and mixed in with the soil sludge atop the Bedrock plates. The lava then hardened the mud with its heat.

Volcanism was off the charts during that year and following as rising lava captured sloshing mud swaths and became one with them into an amalgam of cooling lava rock and warming mud rock. It was heating up and cooking all of the Crust's good soil into "useless rock" (*making all the continents—that "useless rock" that saved all life that disembarked from the Ark*). And since it was holding biomass of all kinds, it reduced it all to either charred coal or liquefied oil or flammable gas (*unless it just fossilized into rock too*). Well, God is going to destroy the world with fire, so He might as well get things ready, right? *All of that flammable stuff on the surface...*

Have you seen the charts and graphs and maps showing massive flows that took place over "several millennia", nay, "millions of years" (*hundreds of millions of years?? Billions of years???*)? They've got them all named, so they must know everything about them, right?

Well, actually this requires a bit of explanation to get a grip on. There were huge flows of mud that got baked into rock. They call the rock "sedimentary". It's all over the world, under every continent. I think that was the good soil of the previous world that was sloshed, tossed, turned, and then baked.

And like I said, that's how fossils are made. But it's also how mountains and other formations are made on the landscape. These massive flows all occurred at roughly the same time—within no more than days of the previous, over a period of about 150 days—as the flood waters were sloshing all of that mud around, which would become the continents when it all comes to rest.

4 | "Earth Looks Old"

Great Unconformity

An example of layers—one mud flow on another.

"Unconformity" is a word used for the divisions in these huge flows of layers (*gaps between flows are unconformities*); the "Great Unconformity" is the bottom divider (*gap*), under which no fossils can be found; only above it. An unconformity is where one mudflow came to rest on top of another. These unconformities are all about fossils and mud-tsunami-flow deposits.

I need to clarify again an earlier statement of mine that there is no erosion present between strata. There are some gaps (*"unconformities"*) in the true layers of the earth under North Americans' feet and others'. These unconformities are the only places that I know of showing signs of erosion between layers. And to me, a "layer" is a giant mudflow with many strata within it. Strata do not show signs of disturbance; layers *are* the disturbance.

But these are showing massive events of extremely rapid erosion flows between layers that were catastrophic and hugely widespread. They are devastating evidence of a worldwide flood event with several sequences of surging mudflows. They formed in rapid succession wherever they exist, globally. Remember I called the grabbing of animals into mud like rolling ants into dough? These flows are the rolls of dough that get piled up in the "oven" to make "raisin cookies".

After all of the dirt finally came to a standstill, much of the topmost layer was washed into the oceans with the receding waters and the high winds. In some higher elevations, huge lakes or seas were formed, only to be loosed later, bringing subsequent flows of much water, with mud and rocks. Look at China and Montana.

Read this quote:

> Many geologists are already aware that there are six thick sequences of fossil-bearing sedimentary strata, known as megasequences, which can be traced right across the North American continent. This was documented five decades ago in 1963 and subsequently verified by numerous observations so that it is now well recognized. In the early 1980s, the American Association of Petroleum Geologists (AAPG) conducted a project in which all the local geologic strata "columns" derived

4 | "Earth Looks Old"

from the mapping of outcrops in local areas, supplemented by drill-hole data, were put on charts to show the sequences of fossil-bearing sedimentary rock layers right across the North American continent.

(Andrew Snelling, AiG answersingenesis.org/the-flood/flood-cataclysm-deposit-uniform-rock-layers)

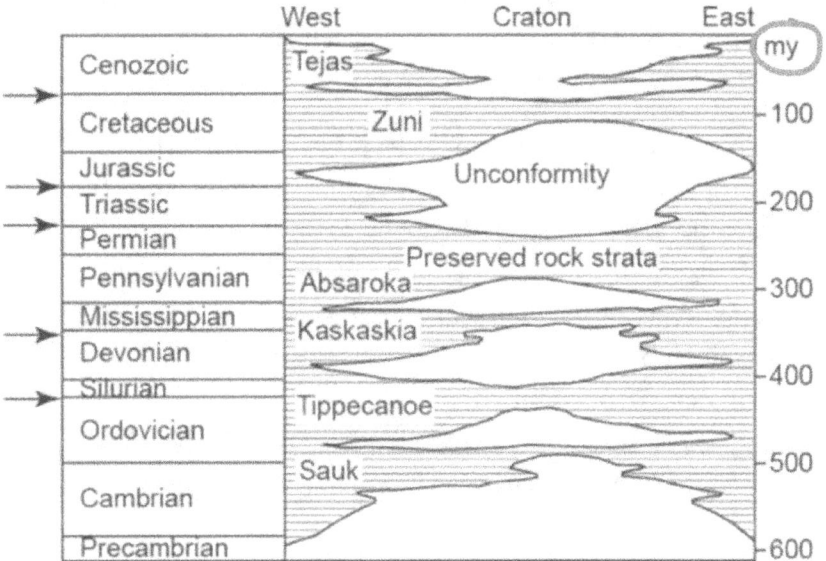

The preserved rock record, consisting of named megasequences, between major unconformities and mass extinctions (arrowed) across the north American continent.

Chart from same article quoted above.

I don't know if you caught that or not, but the major unconformities (*places where mass erosion took place*) match

167

up with levels in the layers where many animals were buried in mud turned to rock.

When animal remains (*bits and pieces, mostly*) and the mudflows in which they are encased are covered up by megatons of more mud layers, and cooked by the high heat of the new lava under and in the Crust below, it's just like a pressure cooker. And we get lovely layers of dead things laid down and covered up in the very same method all around the world at the very same time. Why is this a mystery?

Now that the world knows the processes that cause fossils and fossil fuels to form, we can see that they were present in the world at the Flood, and only during the Flood of the Bible. The conditions were just right during the Flood and never again, before or since, to make fossils and fossil fuels. We call this condition inimitability (*cannot be imitated*). This is true in all of these effects of the worldwide flood and shows up in many ways, such as lava extrusion from the Mantle (*forming the contine*nts), Tectonic Plates formation (*from the breaking up of the Bedrock Layer*), an Ice Age (*due to super saturated skies and severe cold at the poles*), and more.

And so the fossil record agrees with the Bible; not the universities—despite their deceitful cries to the contrary. Their silly explanation of m/billions of years between strata is just completely without any physical support at all in the real world. And the harder they cling to this godless notion, the more foolish they look (*sadly*).

4 | "Earth Looks Old"

Look at the graph again from Andrew Snelling, just under his quote (*above*). It shows "my" at the top of the number column on the right-hand side of the graph. That stands for millions of years, and then the numbers in that column are supposed to be hundreds of millions of years. Well, to show just how "off" they are (*those deep-timers, or old-timers*), if we took their numbers and changed the scale to minutes or hours, not even days (*600 days is longer than the flood lasted*), then that would be closer to reality but still off. Yes, things continued to move for quite a while afterward (*in fact are still in motion*), but I am discussing the formation of the continents. The continents were formed in about 150 days.

Starlight

Here is another argument used by secular believers in Deep Time. The argument goes, "It takes millions of years for the light of stars that are millions of light years away to get here! Therefore, the universe must be old enough for the light to have travelled to Earth from wherever it originated".

Nice one. That's a good one, alright. "It takes millions of years for light to travel millions of light years." (*Millions; billions; whatever.*) It sure sounds good. Now if you could just prove it. But of course, there's no proving that, now is there? It's another one of those theories.

Objection 1

The speed of light may not have been constant then and may not be constant now. Forces can affect the speed and course of light, theoretically. This is unfalsifiable; it's a theory with the same failings as radiometry in a way (*not accounting for the condition at the beginning*).

Objection 2

The Bible says that God created the lights in the sky so that we could see them. But before that, He made light itself.

> Then God said, "Let lights appear in the sky to separate the day from the night. Let them be signs to mark the seasons, days, and years. Let these lights in the sky shine down on the earth." And that is what happened. God made two great lights—the larger one to govern the day, and the smaller one to govern the night. He also made the stars. God set these lights in the sky to light the earth, to govern the day and night, and to separate the light from the darkness. And God saw that it was good.
> And evening passed and morning came, marking the fourth day. (Genesis 1:14-19, NLT)

So according to the Bible, God made the stars, the Sun, and the Moon in the sky on the same day (*the 4th*), within 24 hours. The reason He put them there was to be for us to see them, or see by them. If He created the light for us to see, it would have to be where we can see it on day six, when man was created.

Notice that this is day four in the creation account. The Sun and Moon and Stars didn't exist when God created light on day one! So the **light already existed** between here and there (*wherever the star or other celestial body may be*). There was, theoretically, no reason for the light to travel where it already existed (*all along the path*). Therefore, the light didn't have to travel at all; it was already there where we could see it on day six, when human eyes first saw them. We don't need the light "source" if the light is already there. So He just left the "connections" in the light so we could see it. Logical, yes?

If light travels in a stream, the stream was already flowing when the points of light were put in their places.

If you don't like this explanation, then see how yours isn't going to work either (*since deep time is scientifically impossible*). This is all I have for you at this time. It's a theory; not a revelation. The revelation shows that 6,000 years ago God created all of the lights we can see in the natural universe. Figure it out. I like my explanation.

Every "explanation" given by those who push evil-you-shun and deep time has been solidly debunked over the decades since they came into our limited thinking. And evolutionists have said some of the most condemning things about evolution of all. There is no scientific support for evolution or long ages for Earth, so we can extrapolate this to the rest of the cosmos.

Bonus Support From Bones

If dinosaurs died out *x* million years ago, then why do we see evidence of man and dinosaurs being together? We see images of them in caves and on rocks in the open, we see their footprints with ours (*see photo below*), and their bones often have collagen in them (*quickly deteriorating soft tissue*). And we have numerous accounts of people fighting "dragons", which would be what we call "dinosaurs" today. In the Bible, God talks of dinosaurs with Job (*Leviathan and Behemoth*).

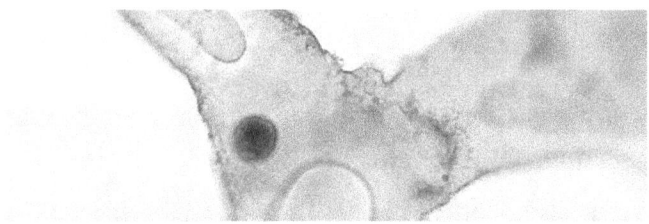

Collagen in Dinosaur bone

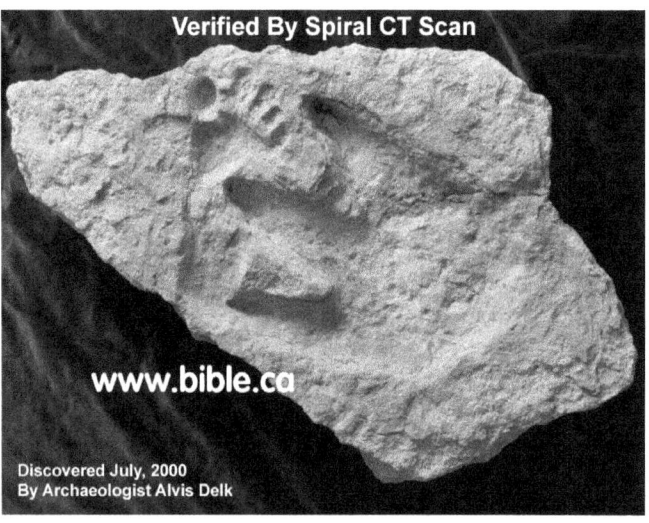

Photo of a human footprint in rock with a dinosaur footprint on top of it.
"Mr. Delk found the loose slab against the bank of the Paluxy River, about one mile north of Dinosaur Valley State Park, [TX]"

4 | "Earth Looks Old"

Group Discussion Questions:

1. Is erosion found between strata or between layers? Pick one.
2. True or False: Mountain ranges were formed during the Flood?
3. What are polystrate fossils?
4. True or False: James Hutton claimed the world is eternal.

Answers:

1. Layers, since they are evidence of huge swaths of mud being deposited by water.
2. True. All mountain ranges were formed during the Flood.
3. Fossils that extend through many strata.
4. True. He made it all up.

Chapter 5 | Radiometric Dating

Let's look at what might be the number one reason why most people reject a Young Earth Biblical Creationist point of view: the so-called dating methods of Radiometric and Radiocarbon Dating. The "science" of deep time. ☻ ♪ ☻

As I understand them (*again, I'm no expert*) is that they both use radiation as a means to determine the age of something. By measuring how much of a radioactive isotope is present in an object (*mineral sample for the Radiometric method, or biomass for the Carbon Dating method*) they say that they can determine the age of the sample they are looking at. Sounds great. Let's do it!

The problems start to arise right away though, as we try to determine the age by counting the isotopes. Those smart people in the labs have some really great measuring devices that they use and I'm sure that they are very accurate too (*my dad was a scientist, working in a national laboratory for 35 years*). But the first problem is going to immediately hit them in the face.

How much of the isotope was present when the rock was formed (*for example*)? I mean, they can count how much is in it now and they can measure the rate, but how much was there when the mineral was formed?

Well, this is a huge problem for the lab. They've got the equipment; they know how to set it up; they know how to run it to count the isotopes; but THEY DON'T KNOW HOW MUCH OF THE ISOTOPE WAS PRESENT IN THE BEGINNING, WHEN THE ROCK WAS FORMED, for example. [face plant]

All of that expensive equipment, all of that expensive training and education, and they are stuck; unable to tell us what we want to know. Why? Because no one had that equipment and training even 100 years ago, much less at the time when that sample was created (*thousands of years ago*). Therefore, the most accurate readings today tell us nothing, or very little, of how much of that isotope was present in the past (*before the equipment and the experts came along*).

This is an objection that radiometric dating has been receiving since its inception. It's not like people didn't complain right away; they did! But the power of the purse prevails. The scam was forced upon us. And I don't expect the enemy (*Satan*) to give up his precious lies that easily. He's the one I put this on, ultimately.

But that's only the first problem with the process; and already you can see that this is fatal to the whole endeavor. If you don't know how much was present at the beginning, then knowing how much is present now is meaningless in a system

5 | Radiometric Dating

of comparison. There's no comparison to make—only hunches and guesses (*smoke and mirrors*).

Unfortunately, they assume that the sample was initially 100% parent isotope and 0% daughter isotope. In other words, when measuring Potassium (K), turning into Argon (Ar), for example, they assume no argon was present in the sample at its creation. How do they know there wasn't any argon in it? They don't. That means that they assume total saturation, conversion, or depletion of the isotope that they are measuring (*however you look at it*). This is not science, kids. It's not even a logical guess! It might, however, be a ruse to fool the populace into believing their story is true (*the one without God in the mix*).

But there are other fatal flaws to the process. One is the speed of decay. They assume a constant rate of change. Why do they assume that? There are forces in the world that can alter the rate of radiometric decay (*speed it up*). Some things, like **heat** or **pressure** (*which are known to have existed everywhere*), can **speed up** the process of decay immensely. But this is not taken into consideration. It's as if they are doing everything they can to get the oldest reading possible, all the time. And they do, thanks to following Hutton's failed paradigm (*with bag placed firmly over head, ignoring reality in front of them*); they use the algorithm, formula, or chart. But really, the decay rate assumption isn't even the core of the problem.

But just the fact that outside forces are not considered into their equations tells me that their equations are meaningless. As it is, no one has any idea at all how much a sample was affected by outside forces and what that did to alter the rate of decay. Not how they're going about it. And no one can claim what they claim with what they know (*I should say, what they don't know*).

So those two unknown factors—outside forces and an unsteady rate of decay—are also enough on their own to nullify any results that might come out of such an activity.

In actuality, according to my model, which no one has ever conceived of, apparently, I can show that irradiation occurred at a specific point of time in the recent past. According to my model, radiation was introduced for the first time in a **huge amount** from the Core of the Earth 4,300 years ago. Before this event, all radiation came from solar or cosmic sources, not the Earth itself! Leaving this out of the equation is why the dating results today are so far off from reality. It isn't just that the methods have flaws, it is that no one has accounted for massive irradiation at a single point in time (*the recent past*). It's like coming upon a pile of walnuts under walnut trees and assuming that all those nuts fell a few at a time over a period of time. What they don't have is the backstory that someone came along and dumped that pile at one time, which makes their assumption not just off, but very off. If it took someone a few minutes to drop of that huge pile of nuts, then saying that they all fell there over a long period of time at the current rate of nut droppage from the trees is just

5 | Radiometric Dating

wrong. Do you see how radiometric dating is completely off now? It isn't that the measurements are bad (*so much, even though they are*), the real problem is the assumption of a **wrong backstory**.

The lab results are in: 3 out of 3 fatal flaws reported. Radiometric data sets are completely and utterly useless to us because we don't know the whole story behind what we can only measure today. The past has not been measured. The effects have not been accounted for. The cause of the overwhelming majority of radiation has not been recognized. Although I can now tell you that it was from the introduction of radioactive lava onto the surface from deep within the Earth.

And that's why we cannot trust anything that comes from a lab in determining the age of Earth using the flawed Radiometric Dating Method. It's just smoke and mirrors. The problem has less to do with the processes and more to do with the backstory used. They assume that the current measurable rate of decay has been happening from the time before time. They do not know that there was an event that placed an enormous amount of radiation into the world in one massive dump.

To really drive this home for you, here is the bottom line to all of this talk of radiometric decay dating: They can measure how much radiation comes to our planet from the Sun and space, called solar and cosmic radiation. They can measure the rate of decay from one element to another, using multiple methods and processes (*and all of them tell the same story*). What they do NOT know is that the world's radiation came to us (*primarily*) from a one-time, inimitable event of a massive irradiation of all life and matter about 4,300 years ago. That introduction of radiation on a massive scale is not accounted for because no one had my model before now.

It isn't their fault, really. I know I come down hard on them and maybe too hard, sorry for that. But I am so frustrated by their continual insistence on getting God out of the picture at literally every turn that I have a very difficult time being patient with their ignorance and attitude. In all fairness though, their ignorance is of a backstory that they have not had access to, ever. And ignorance means not having the data; it is not a statement regarding intent or intelligence (*at all*). My own IQ is nothing to brag about, really!

So can you see that the methods and equipment and even the people are not to blame for the gargantuan error, called radiometric dating? The problem is due to not knowing how it all happened in the first place. My model has that explanation, for the first time in our history (to my knowledge).

If they could bring themselves to admit that it is possible for a massive influx or dump of radiation to have occurred at the Flood of Noah, then perhaps they could get on track and jump off of the slow train of uniformitarianism and onto the high speed train of catastrophism. They'll get farther quicker that way.

5 | Radiometric Dating

I wonder if there's a way to recalibrate the tools, based on the date that we have for known worldwide irradiation (1,656am)? That could be interesting to see.

Carbon Dating

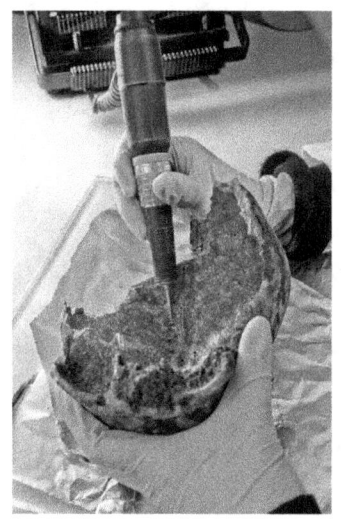

Carbon Dating is little better. It's different because it is counting how much Carbon-14 is there, not how far one (*pure*) element has degraded into another (*radioactive one*). Other than that, it still has similar problems much of the time. This is because they make the same assumptions regarding the sample: an expected initial amount; a steady rate of decay (*disappearance in this case*); and no outside influences. In other words, they guess how much **should** be there, then measure it and subtract the difference, how do they know how much should be there? They use some kind of international reference standard (*a cheat card based on algorithms, I* guess). That's where much of the problem lies.

So the same problem with making assumptions is there. How much was present when the animal or plant died? What forces affected the rate of loss? Was the rate steady or not? Extreme washing, for example, might alter a specimen quite a bit (*making it look older by washing carbon away*). And there was much water in Earth's past (*a whole, worldwide flood of it, actually*).

    Now, Carbon Dating is more accurate than Radiometric Dating (*one element changing into another*) and is only useful when the sample was a living organism, either flora or fauna. However, if someone were to try to use Carbon Dating for a rock, they would have a real problem. If the rock were supposed to be m/billions of years old, then Carbon Dating wouldn't work anyway because carbon only lasts a few thousand years. It deteriorates very quickly, compared to minerals. (*I don't consider carbon to be a mineral. Diamonds, which carbon can be found in, can be made from coal, which is organic. Isn't graphite—another mineral containing carbon—pretty much like coal too?*) So Carbon Dating is not used for the really long times they try to find in minerals; it's the wrong method for that (*as if either is right for anything they do with it*).

    Ironically, these carbon-based materials, whether associated with sedimentary or volcanic rock, are only about 4,300 years old; young enough for carbon dating. The older rocks, if you're into rock ages, are the Tectonic Plates and the Mantle and their crumbs (*those are 6,000 years old*). But everything that makes up the continents, atop the ruptured Mantle and above the Tectonic Plates, is going to be 4,300 years old or newer, made during or since the Flood, out of two substances: lava and/or mud. The 6,000 year-old rocks that make up the Tectonic Plates and the Mantle were made at the creation. Both continental rock forms (*igneous and*

5 | Radiometric Dating

sedimentary) have signs of life, in various forms, caused by the cooking of formerly living things into gas, oil, coal, or fossils, and were deposited at the Flood. But yes, rock is still being made today in many volcanic parts of the earth and sea. If we go to Iceland, we can find rocks that are days old, or being made today! If we swim down to the bottom of the ocean where sea floor spreading is happening, we can see rocks being formed there too.

Anyway, Carbon Dating isn't all that accurate. There are multiple examples on the books of it being off by a factor of hundreds of times more than the reality. And different labs can routinely get different results on the same samples.

You see, we can know exactly how old an animal is if we see it live and die. When it dies we could then measure the C-14 in its carcass to see how accurate the method is. Sometimes it's very close; other times it's very off. I remember hearing that living specimens date at very old ages (*dunno*). Contamination of the sample is likely the biggest factor of the discrepancies.

The bottom line is that Carbon Dating only works so much with biomass and not at all with minerals. The Radiometric Dating Method only could work if we had data from the earliest days of that sample's existence (*or maybe take into account my model*). Otherwise, it is nearly useless much of the time. It is certainly useless in cosmology the way they're using it.

183

Any carbon found in rock samples or fossils would automatically negate millions of years of history for that rock.

Don't trust Carbon Dating too far; it isn't very reliable. It is only marginally valuable at all for dating certain "young" artifacts. It has value, but not much. It can be accurate at times, but not consistently. And pay no attention to the little man behind the curtain, named Al Gore-Rhythm. (*The gremlin is in the math.*)

Instead of trusting in these lab results that tell us nothing, trust the more reliable and actually scientific methods that are available to us. And there are many. In the following chapter are just a few. Deep Time fails any real-world test against it. It's a fabrication with a horrible backstory that does not support the science.

But first I want to show you something from the Bible.

The Contrast

To illustrate the distinction between what the textbooks tell us about life versus what the Creator's Bible tells us, please read the following passage and see if you can respect it. This is not about science; it's about life in general.

> God's promise of entering His rest still stands, so we ought to tremble with fear that some of you might fail to experience it. For this good news—that God has prepared this rest—has been announced to us just as it was to them. But it did them no good because they didn't share the faith of those who listened to God [and

5 | Radiometric Dating

believed]. For only we who believe can enter His rest. As for the others, God said,

"In my anger I took an oath:
'They will never enter my place of rest,'" [Psalm 95:11]

even though this rest has been ready since He made the world. We know it is ready because of the place in the Scriptures where it mentions the seventh day:

"On the seventh day God rested from all His work." [Genesis 2:2]

But in the other passage God said, "They will never enter My place of rest."

So God's rest is there for people to enter, but those who first heard this good news failed to enter because they disobeyed God. So God set another time for entering His rest, and that time is today. God announced this through David much later in the words already quoted:

"Today when you hear his voice,
don't harden your hearts." [Psalm 95:7-8]

Now if Joshua had succeeded in giving them this rest, God would not have spoken about another day of rest still to come. So there is a special rest [Sabbath] still waiting for the people of God. For all who have entered into God's rest have rested from their labors, just as God did after creating the world. So let us do our best to enter that rest. But if we disobey God, as the people of Israel did, we will fall.

For the word of God is alive and powerful. It is sharper than the sharpest two-edged sword, cutting between soul and spirit, between joint and marrow. It exposes our innermost thoughts and desires. Nothing in all creation is hidden from God. Everything is naked and exposed before His eyes, and He is the one to whom we are accountable.

So then, since we have a great High Priest who has entered heaven, Jesus the Son of God, let us hold firmly to what we believe. This High Priest of ours understands our weaknesses, for He faced all of the same testings we do, yet He did not sin. So let us come boldly to the throne of our gracious God. There we will receive His mercy, and we will find grace to help us when we need it most. ~

(Author Unknown, Letter to the Hebrew followers of Christ, chapter four, Holy Bible, NLT)

So the Bible goes far beyond how the world came to be. It goes farther than just saying that there is an intelligent Creator; it introduces us to His pain of how we have jilted, jeered, and judged Him (*and what we've done to each other*).

Misusing radiometry to hide the Creator from the picture (*His picture*) is sin. And it hurts His feelings—His infinite feelings. If you are a believer working in that field, you are

5 | Radiometric Dating

aiding and abetting the enemy—trying to disprove the Bible's youth for Earth.

Funny, how tiny little *we* can hurt great big *Him*. He must really be tenderhearted toward us after all. Try not to break His heart any more than you already have. Believe in Him; not tricks devised by demons and men to pull your attention away from Him and His message. He is real. His message is real. Theirs is not. And their math is funny; not passing the sniff test.

If you are someone who works in the field of radiometry, see what the data looks like under the lens of the timeline given here in this book and flood model. Basically, irradiation occurred about 4,300 years ago for the first time in an explosion of epic, global proportions. Then share your findings. That's how we move forward in science: Take what works; toss what doesn't. And I guarantee that my model works!

Group Discussion Questions:

1. Is the main problem with radiometric dating in the measurements or the initial amounts of radiation?
2. True or False: The irradiation of Earth was a one-time, inimitable event.
3. True or False: Radiometric Dating, used for minerals, counts isotopes and attempts to calculate how long they have been morphing from one element to another.
4. True or False: Carbon-14 dating attempts to count the number of carbon-14 atoms present and then guess how long it's been since the organism quit taking in C14.

Answers:

1. The main problem is the backstory—not realizing that the irradiation occurred in a single, short-lived event 4,300 years ago. Measuring the rate now is moot.
2. True. In the Flood, massive amounts of radiation came up from the Mantle of Earth onto the surface.
3. True. It is about measuring the ratio between two isotopes.
4. True. It is about measuring how much C14 was taken in while the organism lived, then making a calculation.

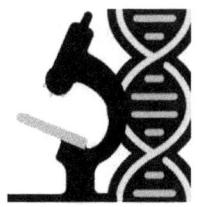

Chapter 6 | Some Scientific Dating Methods

These methods that I'm going to share may not seem like much to the average person, but these are devastating to the Deep Time farce—thinking the world is m/billions of years old—just about as devastating as the fossil record and erosion are. We have data stretching back quite a ways that can confirm certain trends that are observable, quantifiable, and predictable.

Now keep in mind that nothing is absolute and any method can have flaws, and there is no method for proving the age of the Earth scientifically through direct observation. But these methods are simple, uncomplicated, and anyone can see the implications of the observations and data.

The reasons why some scientists push back against these bits of evidence are due to *nonscientific* reasoning; not scientific reasoning. They are still people, after all.

One of these methods is done in a lab, but it uses a much more accurate metric than the failed Deep Time lab methods, with their algorithms and assumptions (*and Huttonian bag placed firmly over head*). This one is actually scientific, meeting the most rigorous of scientific standards, unlike radiometry (*already discussed*).

But first, let's look at a couple of other techniques or data points.

The Sun's Size

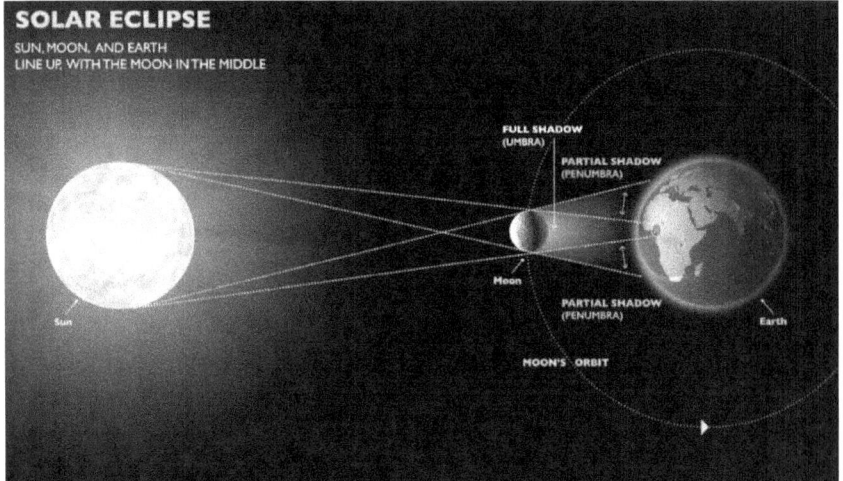

Since our Sun and Moon appear as being the same size from our perspective on Earth (*due to their relative distances from u*s), full eclipses of the Sun by the Moon reveal much to us about the Sun's active surface and its shape and size. These things can reveal more to those who can figure out such things.

One thing that stands out about the Sun that we've been measuring for a very long time now is its changing size. The Sun is shrinking. This is an observable, measurable fact. The rate can be seen to be steady on average (*over the long run*). Yes, it oscillates, but the overall trend has been

6 | Some Scientific Dating Methods

shrinkage. Some say that the oscillations have remained neutral over the eons; meaning that there is no net loss or gain in size. I don't know why they say that.

If we were to have the Sun's shrinking caught on time-lapsed photography, we'd see it growing larger as we went back in time. And guess what? If we were to rewind time just 1 million years, the Sun would be too big for life to exist on Earth. And it would have burned out a long time before 4.5 billion years could have come around too. Oops. Deep time? The Sun says "no".

This one is actually hotly contested by some in the science community. Many do not believe that the Sun is shrinking; while many others do. *And while not everyone agrees that it is shrinking*, most agree that it **will** collapse at some point. (*Go figure.*) ¯_(ツ)_/¯

Comets

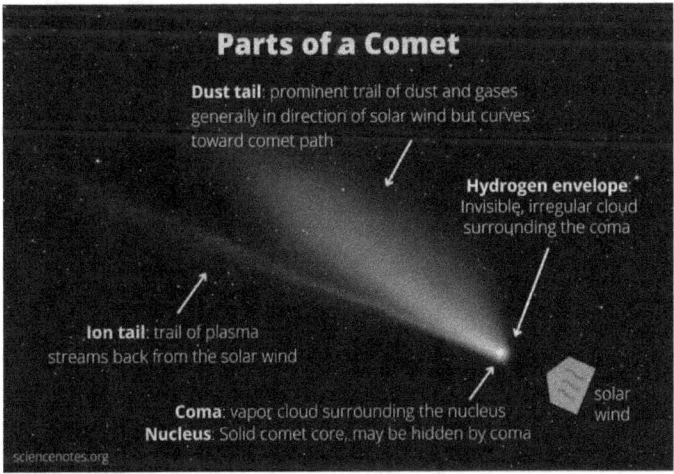

This Broken Planet

I don't know how many comets zip through our solar system, but it's a few. Each comet that we see must be no older than the universe. But how old are they? If we knew that, we might know how old Earth is (*or the universe*).

Well, the thing about comets is like the thing with the Sun; they're shrinking. Comets are losing material all the time as they scream through the cosmos.

If we measure how much material is lost, we might get a rate of degradation, and then see that they are too young to be 4.5 billion years old. In fact, they can't be even a million years old and still have any size at all. There would just be trails of matter strewn about the galaxy; no comets still intact at all.

Some comets, I've heard, are made of ice. I wonder if some of those comets came from the water that was ejected from Earth into space at the explosion of the planet. Just something to ponder. Yes, water and other materials were probably ejected into space. I wouldn't be surprised if there were bits of Earth on the Moon from Earth's devastating explosion 4,300 years ago. Who knows? Maybe all of the comets are earthlings (*little bits of our home planet*). Maybe there's an ice chunk made from Earth water that was ejected into space, or a rock perhaps, that carries some microbes or other lifeforms in it as it travels the cosmos.

6 | Some Scientific Dating Methods

Erosion (again)

As mentioned before while discussing problems with Deep Time, we have evidence on Earth of massive amounts of erosion occurring at the same time all over the planet. Massive amounts of runoff have accumulated on the edge of all seven continents. This phenomenon is called the Continental Shelf.

This is what it looks like when entire continents are subjected to numerous tsunamis in a succession of inundations of mud, ash, sand, silt, clay, and all other forms of soil (*rocks, animal remains, and plants too*).

The entire world was covered with violent flows of water during the deluge. These tsunamis stripped the land of all life and soil, tossing it to the churning waters and slamming it upon the Mantle below, then mixing it in swirls, swift flows, and collisions of opposing forces with rising lava and magma.

It took about 5 months for the continents to be formed from the muddy, soft soil of the original Crust mixing in with he rising lava from the Mantle. This is just the soft soil from the original Crust; the hard Bedrock pieces all quickly went down to (*or into*) the Mantle in a massive chaotic fallout. And then it took about 7 more months for everything to come to a relative standstill, with the top layer of mud getting washed off to form the continental shelf as the waters receded. It took both lava and mud to form the continents. And the lava was still flowing when the waters receded (*it still flows in places, like Hawaii and Iceland*). In my opinion, it really is a miracle that there is any usable soil left in the world at all. I'm certainly not amazed that only 29% of the world is made of dry land, with barely half of that being inhabitable. Antarctica, the Sahara, The Mojave, and other places are not really that inhabitable; not without some serious habitat structuring. Throw in extreme features and weather and there you go. What is that, about 13% usable land? Before the flood, Earth could have been nearly 100% usable from pole to pole with a few seas, lakes, rivers and streams dotted about.

The aftermath of nearly a year of the miles-thick accumulation of flows of mud, ash, and sand, coupled with massive plates of earth being chewed up and slid about upon the ruptured Mantle is nothing less than total devastation of the planet. It is ruined. Bye, bye Old Earth; hello, Middle Earth; come on, New Earth! (*As promised by Jesus.*)

This is what we see: Continental shelves that emphatically scream out about the one-time event we call a "flood". Features on the ripped surface that show how erosion was on a scale unimaginable today (*etching scars into the rocks as other rocks are carried along*). And some of these features are delicate and fragile. People visit such formations as can be found in the Badlands of the Dakotas (*for example*) and marvel at how one rock can precariously balance on top of

6 | Some Scientific Dating Methods

another for so long. Well, we marvel because it is impossible for those formations to last very long at all; just thousands of years, not millions. Even the sides of the Grand Canyon should be smooth and worn way down, compared to reality.

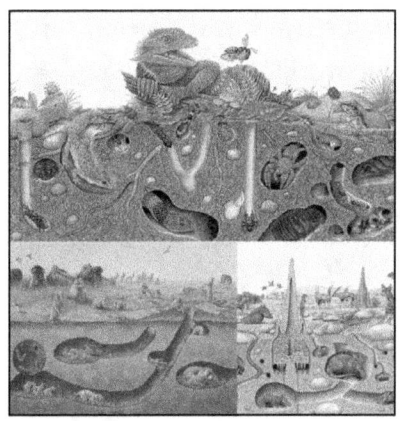

Erosion speaks volumes about the age of the Earth. There's too much of it for there to not have been a flood, and there's not enough erosion for it to have gone on for even tens of thousands of years at the current, mild rate. And the thing about erosion is that it is consistent all across the globe. It speaks more of a worldwide flood than anything else, and everything else confirms it.

Before leaving the topic of erosion, it is important to point out again that there is NO SIGN of erosion between the many "strata" of Earth's Crust, other than the great unconformities, which together scream "FLOOD"! If there were really a million years between "layers" (*a number that is unfathomable for me*) then we would see signs of erosion (*a LOT of it*) between strata, but there is absolutely none of it. Zero. This is terminal for the theory of Deep Time and supportive of a flood. And not just erosion but even the little burrows and tunnels of small animals and insects that scurry on the ground. There is not any (*not any*) evidence of soil disturbance between the strata of sedimentary rock anywhere

This Broken Planet

on Earth, except for those great unconformities between the layers of megaflows. Just dead animals; none of their homes or tracks or burrows. I get sad thinking of it. Of course, the topsoil is full of tracks and signs of life since the flood.

Deep Time has zero support in the real world beneath our feet. I don't see any evidence in the world anywhere of any time beyond a few thousand years. If you can find some, please point it out. Just be honest (*and smart*) about it.

Population Ceiling and DNA

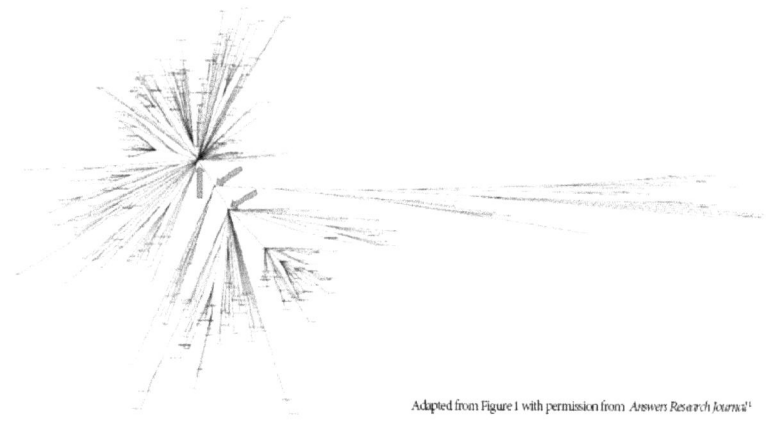

Adapted from Figure 1 with permission from *Answers Research Journal*[1]

An MtDNA genetics graph showing lineages back to Noah's three sons' wives.

If 8 billion is how many people we have in the 4,300-some-odd years since Noah's Flood, how could the world be any older and hold many more people? (*I know this is circular, but it illustrates the point.*) 8 Billion peeps is a big number, but if we had been here any longer than that, the population

6 | Some Scientific Dating Methods

should be in the high billions by now. It's gone up 1 billion in just my lifetime! The parabolic curve we see now in the population data would have started a long time ago.

There just aren't anywhere near enough people in the world for it to be even 20,000 years old. And if the Flood never happened we might have been overrun a long time ago with people (*although that world had a LOT more land than we do now—but no fossil fuels for getting around*).

Sorry, the population just won't let the world be much older than, say, ten thousand years. But I'm going with the 5,986 (*or so*) that are chronicled in the Bible. Let's dive deeper into these two factors of human population growth and migration, and tracking it all scientifically, as DNA can act like a clock looking into the past.

The Low Ceiling of Population

How many grandparents do you have biologically, regardless of whether you knew them or not or if they were even alive or not when you were born? Four, right? (*Two from mom, and two from dad—see how important gender is, kids?*) That means that every person alive today has 4 ancestors in their 2nd generation up. And as you go backward in time, the population seems to grow quickly, because each grandparent has four grandparents and each of those grandparents has four grandparents… But we know that the population actually tends to grow going *forward* in time; more than *backward*. Although, backward numbers are guaranteed; whereas the numbers going forward aren't. So what's going on?

197

Well, we soon find when we follow ancestry backwards that we need to start connecting the family lines between other families, clans, and nations (*Noah's grandkids were siblings and first cousins*). When going backwards in time, the family tree expands quite a bit at first, but then it begins to close back up on itself, joining family lines all around the world until we come back to Noah and his three sons.

You see, at some point, the connections begin to close up going back in time, so that it will meet up with the family lines of Shem, Ham, and Japheth. This is because from those three men all people of the Earth today came into existence. There is no other way that this can go. It's a closed ceiling. The family lines must meet back up somewhere in the past before we get to those three. If this doesn't work, then nothing about Biblical Creationism will. But it does, and it can and has been proven.

There are billions of us between Noah's time and now (*8 billion alive today*). But we can't just keep having the population continue to grow in reverse as we look back in time. For instance, if we were to pick a European country and find a male to trace his Y DNA, we'd soon have too many ancestors for his own country's population. We would need to begin adding in people from the surrounding countries in order to keep up with the count. There would be too few people in the population of that country (*in reality*) just a few hundred years back to support his ancestry. So it becomes necessary, very quickly, to bring in more blood from other people groups.

6 | Some Scientific Dating Methods

And that's how things worked out over time. Different clans became nations, and those nations and their clans interbred with other nations or clans. Through conquest or migration or trade or just being neighbors, couples would come together. Neighboring clans and villages and nations needed to go find mating partners in each other's territories and people at some point.

This means that there was a lot more interbreeding between clans than we might think. But it makes sense when we see this phenomenon. I show this, not to submit proof of a young Earth, but to just show how it all worked out over time. But the proof is there; everything matches up going back in time when we use the short timeline of Young Earth Creationism.

The Y Chromosome

This can be tracked scientifically though, you know. Geneticist, Dr. Nathaniel Jeanson, PhD, has done research into the Y Chromosome and learned some very interesting things for us just recently. I've been presenting his research already in this section.

In his latest book, *Traced: Human DNA's Big Surprise* he presents "an examination of the genetic history of humanity that makes testable

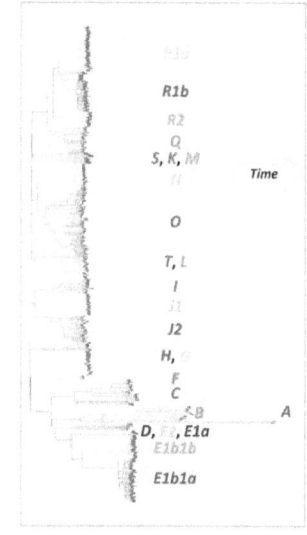

Phylogenetic tree of Y chromosome tracking

predictions from the creationist worldview" *(from his bio on Answers in Genesis).*

Using his technique, which follows the degradation of the male Y chromosome, he can count generations accurately and follow migration paths of groups of people. He just doesn't follow the stretched-out timeline of academia. He tried it, I think, and it didn't work.

Throughout history, following the narrative of those who are in power hasn't always given us the most accurate accounts. Dr. Jeanson's method removes the ability for cover-up or misinformation to replace the truth of what actually happened. He can tell what people groups were involved in the biggest migration events throughout history, and some of the little ones as well.

For example, one of his discoveries led to the validation of an old book that chronicles the history of one of the Native American groups' entry and migration across Canada, North, Central, and South America. The Lakota and Sioux tribes were a part of these migrations.

The book is called the *Wallum Olum* or *Red Record*. It is an example of a historical narrative that was purported to be false; when in fact, it is the true story and the (*mainstream*) alternative is the falsehood, according to Jeanson's research. In it is reference to the biblical flood (*maybe that's why the enemy tried to sweep it under the rug*).

6 | Some Scientific Dating Methods

Dr. Jeanson has learned and can demonstrate that every people group on Earth can genetically trace its origins to Noah and his three sons (*at Babel, I presume, in Mesopotamia*). He says that following the global Y chromosome haplogroup tree is the key to following human history. He tells us that there are around 200 generational steps from Noah to us. Our family tree is a lot shorter than we've been led to believe.

If you believe the false narrative of history taught in schools, then you believe that "Native Americans" came to the Americas from Asia about 15,000 years ago. They then tell us that these people were virtually wiped out by disease brought over in 1492 by Europeans. But Jeanson's genetic research, a scientifically demonstrable method, suggests other migrations occurred much earlier than 1492 from Asia, as well as after.

Jeanson believes that subsequent Asian migrations that occurred prior to the Europeans' mass arrival is actually what wiped out the even earlier migrants from Asia. In other words, it was the "natives" that wiped out the natives (*Asians wiping out Asians*). And that those people who came in later waves did so a lot later than previously believed. In fact, they all came over later than expected. Some of these later "Indian" migrations occurred after the time of Christ; not before. The effect of this is that the timeline is much shorter than the 15,000 years we are told to believe, without evidence. Now we have scientific evidence for early migrations, showing them to be since the Flood of Noah (~4,200 years or so ago). Maybe just a few years after the Babel incident. It's a case of false history trying to cover up the truth. Well, now the truth is falsifiable (*if done right*).

This is an example of a scientifically demonstrable process that debunks a widely held notion that is unfounded scientifically. And it agrees with the Bible completely.

And this process works globally with other civilizations, like the Babylonians, Egyptians, Greeks, and all other ancient civilizations. It is as though every effort has been made by mankind (*or Satan*) to lengthen the timeline at every step of the way from the beginning of time. Is it then any mystery that our timeline is wildly off? And when the Enlightenment crowd got their views pushed into the mainstream consciousness about deep time (*geologically*) it's been the way of the world for 200 years.

This idea that scientists are the smart guys who always do the right thing and search out the truth for truth's sake is just as wrong as saying that people have been making life on Earth better than worse. It's as wrong as saying that a man should be followed just because he says that a higher power says he should be followed, without evidence.

A scientist is a person, with motives and biases. If he makes a statement of backstory or prediction, then that is different than simply counting the beans, so to speak. When scientists begin to wax eloquently about millions of years or other long timeframes, they are not being scientists at that moment; they are being people making stuff up.

6 | Some Scientific Dating Methods

Science is not equivalent to authority. Science is a quest to discover the truth; not make it up. But unfortunately the reverse is what we have been subjected to. People have been calling fact fiction and fiction fact for the whole time. Maybe we should be a bit more skeptical about what we believe and accept as true.

The more I have learned about the Bible; the more I have learned to believe it. I also learn more about the struggle for truth in the world. There are enemies of man that love to spread falsehood, and no one has been able to stop them—and God refuses to; He just lets them tell their lies and will deal with them later, when He judges this world.

If we had a true, detailed, written history of the world (*in addition to the Bible's bits and pieces, I guess*), then we would see that the world is not as old as we have been led to believe the whole time (*either geologically, anthropologically, biologically, or sociologically—heck, all of the so-called "soft" sciences have been contaminated with error and opinion*). But as our data collection gets better, we should be able to drop those tightly held convictions that attempt to force falsehood upon us. Ironic, isn't it?

We've been told how ignorant and backwards the Bible narrative is while being fed narratives that are ten times worse (*millions, even, no, billions!*).

"The Bible's a myth; follow this [fantasy] instead."

It's almost funny. Except for the way that those secular influences allow tyranny and oppression to flourish (*due to removing God's absolute moral law*), which isn't so funny.

Group Discussion Questions:

1. Is the main argument between science and faith about science or the opinions of scientists?
2. What was the book called the *Wallum Olum* or *Red Record* about?
3. Can ancient records be falsified?
4. Can modern interpretations of ancient records be misinterpreted?

Answers:

1. It is the difference of opinion that drives the debates, not the actual data that is correctly collected.
2. The story of early migrants from Asia to the Americas.
3. Absolutely. Many cases exist where a king is exalted falsely, for example.
4. Not just yes, but quite often.

Chapter 7 | Biology

I think it's fascinating to juxtapose the explanations of one biological backstory against those of the other because they're so opposed to any unification with each other at all. And this is shown very clearly in how the basics of their respective views are illustrated.

Evolution

From both creationism and evolutionism, we see imagery used to illustrate the relative concepts. And these illustrations really show us just how opposite these two worldviews are. Both use the very simple and elegant images of trees for their illustrations (*to really get to the root of the issue, I guess*). However, when we see how each view uses the imagery, it becomes quite clear which to follow (*for me, anyway*).

For humanity, evolution seems to not know what to do. On the one hand, it likes to divide up the people into what it calls "races", which would call for many family trees. This is a very foreign concept to the Bible student. As far as I know, there is only one race of humans. We just call us the human race. I call us "family". So evolution wants to see groups of humans coming up separately from each other in various places. [Different Trees]

This Broken Planet

Now, if this story has been abandoned lately, then don't worry about it. But for decades this was their way to show "superiority" between the "races" that they invented, saying that they came up separately from each other around the world.

Worse than that, on the other hand, evolution also uses a single tree to show common ancestry between all living earthlings (*flora & fauna*). Within this device it puts people in with the animals, showing that animals and people are of the same ancestry!

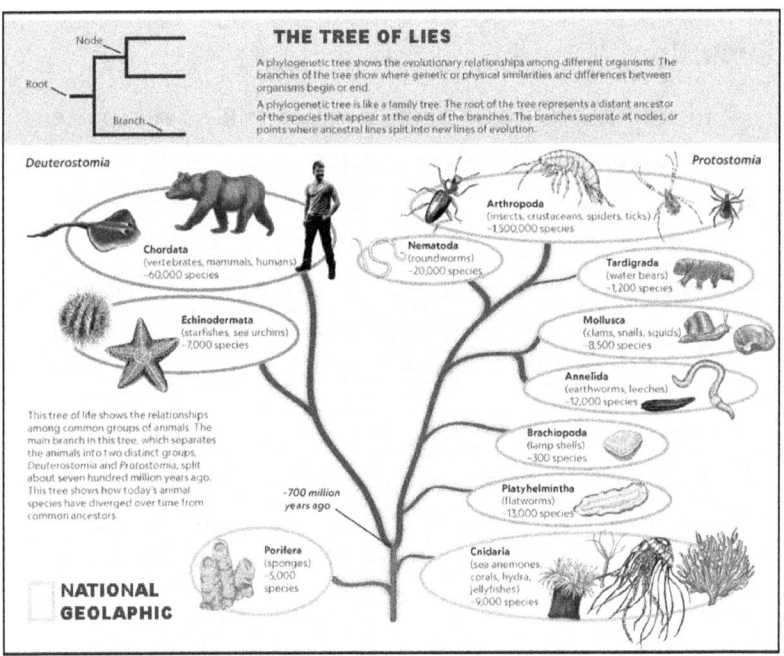

7 | Biology

Wow! What ridiculous nonsense is that?! (*I've been waiting my whole life to proclaim that as publicly as possible—since elementary schoo*l.) I have never for one minute believed such hooey. It's insulting, really. And do you know they include plants in that as well? Look at the image directly above. They're putting us in line with plants and bugs. You know, none of these lifeforms is older than mankind, by more than a few days. This is nothing more than the enemy making fools of anyone who falls for it.

On the one hand, they used this farce to say that certain men aren't men (*equals, peers*) because of something they dug up from an ape someplace, saying that it was an inferior man of some sort. Then they randomly pin that on some poor "race" (*another of their inventions*) so that the others can pick on them. And I'm quite sure, even though I never went to a university, that whoever came up with the whole inferiority thing was not in the inferior category. Hmm.

"Yeah, that's the ticket", he said with a chuckle, "let's just say that these guys are still like their ape grand-pappy, so that we can claim superiority over them".

"Whoa, diabolical."

On the other hand, the creation backstory says that none of the kinds of creatures is related to any other kind; they are all from separate family trees, without a connection, other than all trees being in the same forest (*Earth*) and having the same Creator/Designer. Bible believers follow the concept of equality of all men; regardless of literally anything. You might be better, smarter, faster, yada-yada-yada, but you are not better than any other, as a human being. We are all equal in our value to God. We don't allow bigotry in our presence, do we, Christians? (*Obviously, some have, but that's on them as fallen, depraved individuals.*)

And so it's the image of a forest (*all coming from different origins*) that best explains the origin of the many forms of both flora and fauna; not just one tree that is shared with mankind and all other lifeforms (*what we know of DNA's restrictions disallows it*). Each family has their own family tree; even us.

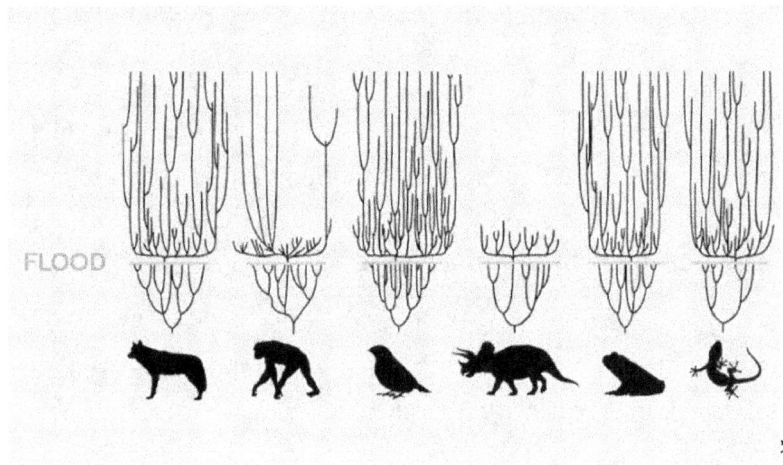

Creation shows each kind (family) growing up independently of each other

In other words: man has always been man; ape has always been ape; dogs are dogs; cats are cats, and never the differing kinds shall merge. KINDS DON'T MIX (like "families").

7 | Biology

Different species of the same biblical "kind" (*family*) can mix and produce much diversity in the kind. I think that God did this for the survival of all kinds; giving them a chance to make it through the various ranges of weather and other harsh conditions of the soon to be hostile planet. As far down as we can dig into the rocks and dirt and ice, we will see this evident truth in the fossil record—a trilobite is a trilobite.

Don't like trilobites? Okay, wherever we dig, a man is a man. A squid is a squid. And on and on it goes. It is a universal truth of biology. And it shows a steady amount of distinction everywhere; for all time, that we can sift through. From the very deepest to the shallowest, wherever we dig them up, kinds are kinds and they are the same as they are now (*mostly*).

"Then where did the dumb tree come from that shows a mythical way of seeing life (*putting all animals on the same family tree of genetics*)?"
"Well, it was made up, of course."

The only reason why I can imagine such a lie being manufactured would be to remove the Creator from the backstory. Why would they do that? I guess they just didn't like the truth. Maybe He scared them (*and rightly so*). Who knows? My fear is that they are just pawns of the enemy. Whatever demons can do to get people to not think about God, they will do; whatever it takes. They never slow down or take a rest. They are invisible and sometimes play to our senses for favorable input that leads to an unfavorable outcome (*if we fall for it*).

In the real world, we know that genetics does not allow one animal to change into another animal. It just doesn't allow that kind of change to happen. We may be able to track differences in bird features on a remote island (*like their beak shapes*), but we still can't say what the bird was supposed to have been before it was a bird. There are limits imposed into the genetic sequencing that only allow changes to go so far and no more. Variation within a kind is everywhere. Kinds becoming different kinds has never been seen anywhere on Earth. Ever. It is superstition to believe so. And who's to say that Darwin's finches didn't already have different beaks onboard the Ark (*there were 7 pairs onboard, after all*)?

If you want to call this ability to change slightly "adaptation", be my guest. If you want to go out of the realm of possibility and say that the same mechanism can allow a slug to become a skink, then we cannot agree. And I might laugh at the absurdity of it (*like I did in elementary school*). Even a child knows that such a thing is ridiculous (*at least, I did—instantly*). And we have proven such in labs across the world in this millennium and the previous one as well. We have discovered limits to changes.

Biology does NOT allow for evolution between animals of differing families (*kinds*). It only allows for slight differences in the lifeforms (*species*), while keeping their unique traits unique for their kind. Mostly, mutations are destructive to the

7 | Biology

lifeform's health and continuance. To think that there could be an upward trend in DNA malformations is absurd. By "upward", I mean "positive", "helpful", "healthy", "good", etc. It is not scientific to believe that the trend has been positive when mutations occur; only changes that are within the scope of the family traits are good. See Entropy. The overall effect then is a lot of sameness throughout the species and the kinds.

For instance, all cats are catlike. All dogs are doglike. Their unique traits remain their own. This doesn't preclude similarities between families; those are there too. Both dogs and cats have tails, and paws, and walk on all four legs primarily. But their differences, the things that make them catlike or doglike, never go away from the family. It's a family trait for a reason.

Books, upon books, upon books have been written showing how evolutionary thinking is completely wrong for life on Earth (*including biologically*) in every imaginable way (*many written by evolutionists*). Yet, the powerful schools keep fighting back and they keep winning the battle over keeping the Bible (*and truth*) out of our schools (*"Separation of Church and State" in America is BS*). Some nasty people have even used that excuse to keep little Johnny from praying over his questionable lunchmeat and gnat-garnished gelatin in the school cafeteria. Poor Johnny needed to pray over that "food".

The truth is out there. We just need to take the bag off the heads of the scientists so that they can do science without stupid restrictions and guidelines (*like, "the present informs the past"; or the world is eternal...billions...millions...*). If you are a scientist, see how my Broken Planet Model can inform your discoveries. And don't let "them" bully you into not studying it or writing about it. Entropy is real. Factor-in God's story with your outlook, you'll get farther, faster.

Evil-You-Shun

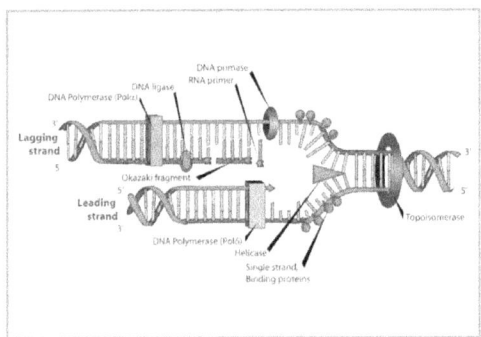

There is great capacity for good in biology: fighting disease or making life better through understanding and caring for the life around us and in us. But there is also great capacity for evil to be done, medically. And we are seeing it already around the world.

Factories and labs like in Wuhan, China produce a great quantity and variety of biological weapons. Laboratories all around the world, here and everywhere, are doing things with DNA that they should not be doing—like combining human DNA with that of animals. This is evil of the worst kind. We must fight against the labs doing such evil. Only, the politicians are using it to control the population of the world (*they tried to take out billions of us just a few years ago*).

7 | Biology

"But it's in the name of 'science' and 'discovery'" you might protest.

"No, it's in the name of Satan". I would reply.

Genetically Modified Organisms (*GMOs*), whether flora or fauna, are creating real problems in the health of humanity on all levels. The labs of food companies only care about speed of production and appearance in the grocery store. Actual nutrient content or health isn't even on their plate.

Worse than that, strange animal remains have been said to appear in the waste stream of labs all around the world, including America. Animal remains (*obvious abominations of attempts to combine two creatures that should not be*) are allegedly beginning to emerge in sewers and waterways all over.

They aren't viable lifeforms (*that I know of*), but if true, the scientists are trying to do things they ought not to do. I wouldn't be surprised at all if I heard that a new hybrid of human and something else were created somehow in a lab. Or if they announced an alien race that has come to the world (*but it wouldn't really be what they say; it would be a lie that is as manufactured as the lifeform they produce.*) Oh, but the capacity of that one for social control. (*It's really scary.*) Or imagine demons, posing as aliens (*more so than now*), running the world outright. "We are in charge now!" And they really, actually, literally are **extra** (*not of*) **terrestrial** (*this world*).

Evil-You-Shun has been used to justify the killing of people whom an evil dictator said were inferior to him because they hadn't evolved as much as him and people who looked like him—as if there's a valid excuse. (*Hitler was an example of this behavior, even though he had no Aryan features, himself.*) He justified the mass murder of millions of people using evolution (*as one excuse of many*). Again, there is no excuse.

At the heart of it, evolution is evil because it is a lie that attempts to remove God and His truth from human consciousness. And when that happens, all hell can and usually does break loose. It's about the worst kind of deceit we can come up with. I mean, look at how bad people are *with* His influence; imagine how bad we'd be *without* Him at all! Well, we're just about there as a planet. And evolution, with its deep time farce, has been ushering-in just that kind of world at a very quick clip. You can remove God's rules; but in the process you'll remove His blessings too. That's just how He works.

Let's quit talking about how bad religion has been in our society and just drop it. Stay out of religion. But don't take up another system of bondage in the process (*like the schools wield—schools & cults; what's the diff?*).

7 | Biology

Instead, simply recognize Who the Creator is and thank Him and worship Him for His goodness (*you'll have to read the Bible for that info*). Ignore the bad parts of human behavior and religious inventions. Just focus on God and His goodness toward us. Not in a religious, got to impress with goodness, kind of way; but in humility, honesty, fear, and a genuine desire to be with Him as an obedient child. This can only come through the sacrifice of Jesus on the Cross—start there, since He did the work so you don't have to (*and you couldn't anyway*.) And don't stop meeting together to worship and praise Him and care for each other's needs in brotherly love.

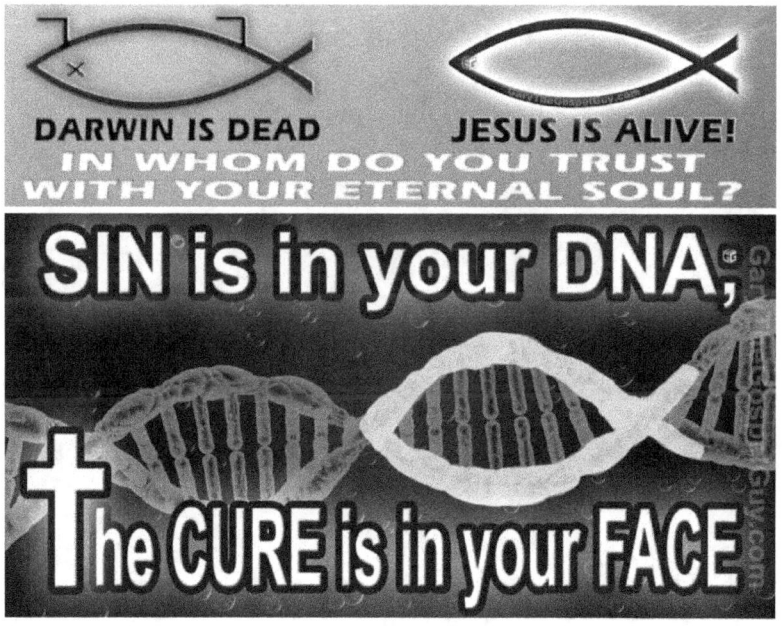

Group Discussion Questions:

1. Does DNA support or refute evolution?
2. How does DNA refute evolution?
3. Can one kind of animal turn into another?
4. Is life better illustrated with a single tree or a forest of trees?

Answers:

1. It refutes it, due to the stringent limits it places on reproduction.
2. Offspring have the same characteristics as their parents. Always.
3. No. Dogs are dogs and cats are cats. Never the two shall meet.
4. A forest of trees, showing separate lines of ancestry.

Chapter 8 | The Promise

"What's all bad news and no good news at all?"

"Depressing."

This isn't a "doom and gloom" message; it's a sober look at where we are, how we got here, and what's coming in the future (*all based on the Bible, with a personal revelation from God, and observation of the Earth*). And we have hope (*I know I do*). Yes, there is the threat looming over our heads from the Creator that we need to pay attention to Him (*or else*). But He is offering us peace and happiness that is on a level largely unknown to us at this time on this planet. (*I experienced a brief moment of this and it is to die for; but that's another vision.*) Let your response be a godly fear of Him and His justice that demands payment. My focus is on His joy living in me.

On the one hand, we have an offer of peace and fulfillment; on the other, we have a guarantee of destruction and misery. "Hmm, that's a tough one; overwhelming misery or overwhelming joy…". But because He has shown us what He will do if we are evil toward Him and each other (*destruction in misery*), and that it is frightening to us (*rightly*

217

so), many people end up judging the Judge instead of heeding His warnings. This is unfortunate. Please don't be that person. It's okay to fear Him. He's frightening. Just don't judge Him. That's His aria to sing.

   Instead, read what He wants to do with His people and see if you want in on that or not. If you do, just listen to Him and obey. Belief will become stronger as you seek His will, His face, His presence and His ways. Just don't expect to impress Him with your "goodness". And don't expect Him to "pay up" what you might think He owes you (*through religion or other such activity*).

> "Look! I am creating new heavens and a new earth, and no one will even think about the old ones anymore. Be glad; rejoice forever in my creation! And look! I will create [New] Jerusalem as a place of happiness. Her people will be a source of joy. I will rejoice over [New] Jerusalem and delight in my people. And the sound of weeping and crying will be heard in it no more.
>
> "No longer will babies die when only a few days old. No longer will adults die before they have lived a full life. No longer will people be considered old at one hundred! Only the cursed will die that young! In those days people will live in the houses they build and eat the fruit of their own vineyards.
>
> "Unlike the past, invaders will not take their houses and confiscate their vineyards. For my people will live as long as trees, and my chosen ones will have time to enjoy their hard-won gains. They will not work in vain,

and their children will not be doomed to misfortune. For they are people blessed by the LORD, and their children, too, will be blessed. I will answer them before they even call to me. While they are still talking about their needs, I will go ahead and answer their prayers! The wolf and the lamb will feed together. The lion will eat hay like a cow. But the snakes will eat dust. In those days no one will be hurt or destroyed on my holy mountain. I, the LORD, have spoken!"
(Isaiah 65:17-25, NLT)

(Poor snake. Still not forgiven for letting Satan in, I guess.)

And that passage from Isaiah is describing life for the mortals in the New Earth, those called "The Nations". The saints will enjoy a much higher kind of fulfillment, joy, and purpose. The ones you just read about will be under the rule of the saints and Jesus. It is a lower form of humanity than what the saints will inherit for all eternity!

On Jesus and Destruction

Jesus has the authority, the right, to destroy that which He has made. He also has the means, the will, and the power to destroy. We should not be testing His resolve, His will, His patience, or His grace and mercy. Think of what He has done to rebellious mankind already (*His actual track record*). He has always done what He set out to do and He was never sidetracked or distracted from His task.

- Death was given to humanity at the Fall for eating the wrong fruit (*disobedience*)
- He broke the planet because mankind was being evil with each other and Him; millions were killed (*all they wanted was sex and violence—I'm sure that drugs and music may have been involved as well*)
- The entire Jewish history with God is chock-full of Him punishing them for their unfaithfulness to Him, while praising Abraham for his faith
- He has promised His followers trouble in this lifetime, which they are (*and have been*) receiving
- Early on, He prophesied His destruction of Earth at the end
- We can see This Broken Planet and know that He broke it because of us.

Satan isn't the only one able to destroy. But he is evil because he destroys what is not his to destroy. He destroys for his own twisted purposes. His destructive acts are evil tantrums. Jesus can also destroy. In fact, Jesus has the power, the right, and the will to destroy Satan, but Satan can't even touch Jesus anymore. He tried and failed.

Jesus only destroys what the Father wants Him to destroy. Destruction doesn't bring Him joy; but it does bring Him and His people peace, which eventually does lead to joy. He brings all things under His dominion and then will hand it all over to the Father when He is ready (*full of joy*). He will eventually imprison Satan in a Lake of Fire (*or lava, if you will*) forever and ever.

Satan's power is limited, but Jesus' power is unlimited. Jesus promises to destroy Satan, this universe, hell, and heaven after these coming thousand years after His soon return have passed. He will then create a New Heaven and Earth. Real life will begin. (*This world is just the interview process.*)

Religion

I've heard many people over the years talk about their dislike for "organized religion". I don't really know what they mean by that, since all religion is organized. That's kind of the point of it, isn't it? To get together and talk about the same thing from the same viewpoint and share that with each other, whatever the thing is called, over a potluck? I think that's a part of it. Actually, what I think they mean is that they just follow their own made up beliefs—and they keep them secret to avoid criticism.

Here's my take on this thing called "religion" (*not that you asked for it or care, but see if I can get this right*):

> Religion says that God owes me for doing what I needed to do. It's like a paycheck; "I did the work, now give me my pay." ~ (me)

Can you think of a single religion that doesn't incorporate some level of effort that it requires of you? And let's include Christianity and Judaism in the count, too. **Religion tells God that He owes us.** If you think that God owes you good things because you are a good person, remember what Isaiah said; all of our best behavior is disgusting to Him (*see Isaiah 64:6 quoted below under Relationship*). God owes nothing to anyone ever conceived. Religion tries to avoid that reality and make Him owe us. This is why I hate religion; it's wrong.

I personally think that one of the biggest issues facing Christianity is having Satan in the church. I don't mean that everyone is demon-possessed. I mean that he's right there with you, sitting in the pews (*or floating above everyone's head, if you like*), trying to influence everyone as much as possible.

I think that he enjoys being the pastor, or the worship leader, or a cardinal, or a pope. Or maybe he's that older, heavier woman, who just loves to get into everyone's business and shake things up (*and not just with too much perfume, either*). But with a smile and a hug, of course.

What am I talking about? It's that thing that happens in everyone's head (*self included*), when they have a thought come to mind and it could either be from the dark side (*evil*) or the light side (*good*). It's kind of like the cartoon of a red, horned devil on one shoulder and the white winged, haloed angel on the other; both whispering in the toon's ear.

Only, this happens at lightning speed with each and every thought that comes to mind. I know that my own thought-per-second count can be pretty high sometimes. I prefer when it is more like thoughts per minute, or per hour. But that doesn't happen much with me (*they call it ADHD*).

[Or emotions. When I say *thoughts*, it could apply to the *emotions* as well—same brain (*or soul*).]

And if I don't make sure that I'm dwelling on the good thoughts (*or feelings*), then the bad are what I dwell on (*sometimes; I do get neutral thoughts, like when working on a task*). And the longer I dwell on the bad thoughts, the worse my mood becomes and the more likely I am to do or say something as foul as my thoughts. That is when the enemy is in charge and I have failed in standing up to him. When the pastor gives that evil thought a place to live in his heart, then it is not him who is delivering the sermons. I share this because I'm just like you in this. Watch your thoughts and emotions, kids.

The funny thing about people is that we can be both good and bad at the same time. In fact, we can't help it. And religion comes in and takes advantage of that trait in us and uses guilt, or reward, or whatever works to get us to engage in something that is futile in making us right with the Creator. Or worse, we just start picking on other people. If you are a Christian who picks on others, you are not living as a child of God. If you look down on those who don't live up to your code, you are not living as a child of God. If you think that you can do no wrong, you are not living as a child of God.

Any person at any time has the capacity for good or evil. Well, infants get a pass for now (*from me*), but toddlers don't. They're evil little beings; those toddlers. You have to watch them every second! Be tough on them now or they'll rule you in a few years! I'm not kidding. But don't get me going about parenting.

Babies in the womb are technically just as guilty as the rest of us, in that they have the very same Depraved Nature of Adam—that thing in us that is not removable—which keeps us from having fellowship with God. I just give them a pass because I can. But I'm not God. I'm not the Judge.

In a nutshell, religion is a device designed to keep control over the masses. And in the past, Christianity has been as horrible as many of the others and worse than some. Did you know that the statistics between those households involved in religion (*including Christianity and Judaism*) are the same as the larger society in all categories? Murder, rape, theft, incest, abortion, addictions, and single parent households—any of the plagues on society and family—are just as prevalent in the households of the religious as in the other households. That tells me that religion doesn't work in doing what its proponents often say that it will do.

And look at science, the same kind of thinking that ruined Christianity is rife within that community and its disciplines as well. If you don't see it, then you're a part of the problem.

Evolution and Deep Time are not scientific; they are beliefs (*ruses, really*) that counteract a God-based paradigm. They are part of a response to a religion (*Christianity*) that has been demonically influenced and hurtful toward society in many ways at various times. It isn't the story that is out of whack, the Bible is just fine, it's the crazy ideas and behavior from the pew-sitters, priests, popes, and theologians that's out of hand—along with their godless lifestyles. Poor choices, second-by-second are what we have seen from the Church for the last two millennia. But then, we're just fallible people too. Right?

> If you claim to be religious but don't control your tongue, you are fooling yourself, and your religion is worthless. (James 1:26, NLT)
>
> Why, then, do you judge your brother? Or why do you belittle your brother? For we will all stand before God's judgment seat. (Romans 14:10, Berean Standard Bible)
>
> You may think you can condemn such people, but you are just as bad, and you have no excuse! When you say they are wicked and should be punished, you are condemning yourself, for you who judge others do these very same things. (Romans 2:1, NLT)

> "Do not judge, or you will be judged. For with the same judgment you pronounce, you will be judged; and with the measure you use, it will be measured to you. (Matthew 7:1-2, Berean Standard Bible)

It isn't your "being good" that God empowers your salvation with; it's the faith in what God has already done that pleases Him. He wants you to seek after Him, just to be with Him; not to prove yourself good enough, because you aren't—nobody is. No one. Just show some loyalty, you know? Listen. Obey. Pray daily. Read His Word . . . .

I think that having an imaginary long time between God and us, in theory, is like a buffer to some. It makes Him seem like He's farther away than He really is; as if our judgment will be less likely, perhaps. I think that many people in the world since Babel have been so scared of God's judgment, or reviled by the force with which He deals with sin, that they have made up other stories to placate their fears, or those of their children, when the tale is told. I don't know. I'm just trying to figure out why people reject a good God. So far, this is all I've come up with: some people don't think that He's good. Weird. When they try to hide under satanic lies that they've latched onto, as if their faith in the wrong thing is as effective as our God-given faith in the Real Thing (*Christ*), then Jesus (*The Judge*) will not be amused, I would imagine. There is also the thing about some people resenting His judgment over their choices and desires.

Relationship

Let me see if I can really nail down the biblical side of this for you, if you still have a hard time seeing how this all works out. I know it can be confusing when religion is involved.

There is only the one God. We know this from the Bible, which, as I've been trying to demonstrate here is trustworthy in all things it discusses, including the creation. (*If you don't agree, prove it wrong.*)

He revealed Himself to the post-Flood, post-Babel world through people He calls His prophets; people through whom He speaks or communicates His truths.

One of those prophets was named Isaiah. He is considered a major prophet because he was so prolific and impactful. In his book of prophecy, he wrote the following:

> We are all infected and impure with sin. When we display our righteous deeds, they are nothing but filthy rags. Like autumn leaves, we wither and fall, and our sins sweep us away like the wind. (Isaiah 64:6, NLT)

If you aren't sure, "righteous deeds" means the very best behavior possible.

In this one quote, God is basically removing the point of having any religion toward Him at all *that seeks to use human effort to please Him*. Why? Because according to Him, the very best we will ever be in His sight is as disgusting as a used tampon. Sorry, but there it is. The Hebrew term used by Isaiah for "filthy rags" could indicate a menstrual rag. So, the best we can hope for is to not gross Him out with our disgusting behavior and character (*the best we can come up with*).

His perfection is unattainable for us on our own. We are never going to be perfect, as we exist now. We have to be transformed somehow. But hang on to that thought; I'll get back to that.

First, I want to talk about this God that people don't seem to understand; the One of the Bible.

Firstly, I don't see why they use ignorance of knowing everything about Him as an excuse for not pursuing Him. I mean, agnosticism is just willful ignorance in my eyes. But then, my eyes have been opened to the spiritual reality of my Creator—unlike most people in the world, I guess. And until a person's eyes are opened, they won't see it. (*You have to believe it to see it*.) So, forgive my seeming insensitivity to the person who really wants to know God but hasn't met Him yet.

Well, I actually believe that if a person wants to know God then He will introduce Himself to that person. So let me see if I can inspire you to see why having a good relationship with God is better than having a bad one. And everyone has some kind of relationship with God already. The quality or character of it is what matters here.

And our relationships with each other matter as well. Having common trust in God in this world, that is hostile to that trust, makes fellowship in that trust all the sweeter. God works through these relationships (*when we let Him*) to strengthen and encourage us to continue in the fight against the enemy. And the enemy is internal as well as external, for me. And the domestic enemy can always do more damage than the foreign.

Who Is God In the Bible?

"I AM THAT I AM" is what He told Moses, when Moses asked Him for His name directly. "Tell them that 'I AM' has sent you." That Hebrew name is transliterated into English as "Yahweh". So one name for God is YAHWEH. Jews tend to not say the whole name out of reverence for its holiness and they write it "YHWH" (*but in Hebraic letters*).

The Bible describes His attributes and character as it goes along. I think that every book of the Bible has something to say about Him and His character or His ways.

When I was a child, I had received a bright red Bible at a church youth event. (*My older sister and her friends conspired and cheated and I ended up with the prize for the most buttons or pins*). I might have been in first grade, or close to it. It was in the King James Version. It took me a little while to understand it (*still trying to get there, actually*). The King James Version (*KJV*) is not easy!

What I loved about it was that the words of Jesus were in RED. So I remember just reading the red parts when I could. Over and over. That's when I fell in love with Him. In grade school and getting what I could out of the KJV Bible. I'm not a big fan of the KJV today. I recommend another, more connectable version. It's hard to relate with words that are out of date. It's good for poetic verses though.

My personal favorite right now is the NLT. But it isn't meant for all kinds of study; it's best for casual or regular reading (*it works well for this kind of project*). A person should read all of the versions of a particularly important, touching or appropriate verse, just to see what the translators were trying to convey as well as the original author. I also recommend seeing the original languages as well (*Greek or Hebrew, primarily*). Have and use a tool for translation away from any particular version. Interlinear Bibles are a must for me. Concordances are great too.

Avoid these Bibles: The New World Translation; The Mirror Bible; The Message Bible, or just The Message; The Living Bible; The Passion Translation; and The Reader's Digest Bible.

Anyway, back to Jesus. Yes, Jesus. He was shown to be God in human flesh many centuries earlier by that same prophet that I quoted earlier—Isaiah.

If you will read the 53rd chapter of his prophecy, you will see him describing Jesus and His substitutional death on the cross for all of us, whether we asked for it or not.

The grueling, torturous, slow death that He suffered is one for the record books—and it is. His life and death were so momentous that we began counting our years from His birth on (*more or less*). If you forgot; AD *anno domini* means year of His dominion. (*That's why the college boys wanted to change it to CE, Common Era, and BCE, Before Common Era.*) I prefer AM, *anno mundi*, because it helps me count the days to 6,000 (*the magic number*). Here is just one prophecy about the most important act and human being for all mankind:

Isaiah 53

In the New Living Translation

> Who has believed our message? To whom has the LORD revealed his powerful arm?
>
> My servant grew up in the LORD's presence like a tender green shoot, like a root in dry ground. There was nothing beautiful or majestic about his appearance, nothing to attract us to him.
>
> He was despised and rejected—a man of sorrows, acquainted with deepest grief. We turned our backs on him and looked the other way. He was despised, and we did not care.

Yet it was our weaknesses he carried; it was our sorrows that weighed him down. And we thought his troubles were a punishment from God, a punishment for his own sins!

But he was pierced for our rebellion, crushed for our sins. He was beaten so we could be whole. He was whipped so we could be healed.

All of us, like sheep, have strayed away. We have left God's paths to follow our own. Yet the LORD laid on him the sins of us all.

He was oppressed and treated harshly, yet he never said a word. He was led like a lamb to the slaughter. And as a sheep is silent before the shearers, he did not open his mouth.

Unjustly condemned, he was led away. No one cared that he died without descendants, that his life was cut short in midstream. But he was struck down for the rebellion of my people.

He had done no wrong and had never deceived anyone. But he was buried like a criminal; he was put in a rich man's grave.

But it was the LORD's good plan to crush him and cause him grief. Yet when his life is made an offering for sin, he will have many descendants. He will enjoy a long life, and the LORD's good plan will prosper in his hands.

When he sees all that is accomplished by his anguish, he will be satisfied. And because of his experience, my righteous servant will make it possible for many to be counted righteous, for he will bear all their sins.

I will give him the honors of a victorious soldier, because he exposed himself to death. He was counted among the rebels. He bore the sins of many and interceded for rebels.

That isn't myth or fantasy; it's a realistic narration of an event that would happen 700 years after the time that it was penned—the life, death, and burial of Jesus. In fact, if all of the prophecies about Jesus were shown to you it would be quite obvious Who He was in the eyes of the prophets.

The Torah is not myth or fiction. Neither is the Tanach. The only poetic books are obviously poetic (*like Psalms and Proverbs*). Most are narrative of actual history (*yet to occur or not*). And the prophecies in the Tanach are all either perfectly carried out or they have yet to be fulfilled. There is not one prophecy in the Bible that can be confirmed to be false, if it says it's true. *I think it does mention a false prophecy or two; not sure.*

Jesus said this about Himself:

"Don't let your hearts be troubled. Trust in God, and trust also in me. There is more than enough room in my Father's home. If this were not so, would I have told you that I am going to prepare a place for you? When everything is ready, I will come and get you, so that you will always be with me where I am. And you know the way to where I am going."

"No, we don't know, Lord," Thomas said. "We have no idea where you are going, so how can we know the way?"

Jesus told him, "I am the way, the truth, and the life. No one can come to the Father except through me. If you had really known me, you would know who my Father is. From now on, you do know him and have seen him!"

Philip said, "Lord, show us the Father, and we will be satisfied."

Jesus replied, "Have I been with you all this time, Philip, and yet you still don't know who I am? Anyone who has seen me has seen the Father! So why are you asking me to show him to you? Don't you believe that I am in the Father and the Father is in me? The words I speak are not my own, but my Father who lives in me does his work through me. Just believe that I am in the Father and the Father is in me. Or at least believe because of the work you have seen me do."
(John 14:1-11, NLT)

Even His closest friends had their misunderstandings about Him. But now we know that He is the physical representation of the great and glorious Yahweh.

Paul, writing to the church in Corinth, in the New Testament, tells us that Jesus is the One Who created all that is.

There may be so-called gods both in heaven and on earth, and some people actually worship many gods and many lords. But we know that there is only one God, the Father, who created everything, and we live for him. And there is only one Lord, Jesus Christ, through whom God made everything and through whom we have been given life. (1 Corinthians 8:5-6, NLT)

Jesus is called "God" by God in the book of Hebrews:

Long ago God spoke many times and in many ways to our ancestors through the prophets.

And now in these final days, he has spoken to us through his Son. God promised everything to the Son as an inheritance, and through the Son he created the universe.

The Son radiates God's own glory and expresses the very character of God, and he sustains everything by the mighty power of his command. When he had cleansed us from our sins, he sat down in the place of honor at the right hand of the majestic God in heaven.

This shows that the Son is far greater than the angels, just as the name God gave him is greater than their names.

For God never said to any angel what he said to Jesus: "You are my Son. Today I have become your Father." God also said, "I will be his Father, and he will be my Son."

> And when he brought his supreme Son into the world, God said, "Let all of God's angels worship him."
>
> Regarding the angels, he says, "He sends his angels like the winds, his servants like flames of fire."
>
> But to the Son he says, "Your throne, O God, endures forever and ever. You rule with a scepter of justice.
>
> "You love justice and hate evil. Therefore, O God, your God has anointed you, pouring out the oil of joy on you more than on anyone else."
>
> He also says to the Son, "In the beginning, Lord, you laid the foundation of the earth and made the heavens with your hands. They will perish, but you remain forever. They will wear out like old clothing. You will fold them up like a cloak and discard them like old clothing. But you are always the same; you will live forever."
>
> And God never said to any of the angels, "Sit in the place of honor at my right hand until I humble your enemies, making them a footstool under your feet."
>
> Therefore, angels are only servants—spirits sent to care for people who will inherit salvation.
> (Hebrews 1, NLT)

And Jesus is very clearly shown to be God in human flesh in the Bible. Sorry, Jehovah's Witnesses, you're wrong. God lifted up Jesus to His level (*almost*), with the Holy Spirit in them both. Soon Jesus will lift saints up to His level (*almost*), with the Holy Spirit in us too. It's about the most beautiful thing in all, being one with God and owning your life.

> Watch out for those dogs, those people who do evil, those mutilators who say you must be circumcised to be saved.
> For we who worship by the Spirit of God are the ones who are truly circumcised. We rely on what Christ Jesus has done for us. We put no confidence in human effort. (Philippians 3:2-3, NLT)

> Dear brothers and sisters, pattern your lives after mine, and learn from those who follow our example. For I have told you often before, and I say it again with tears in my eyes, that there are many whose conduct shows they are really enemies of the cross of Christ. But we are citizens of heaven, where the Lord Jesus Christ lives. And we are eagerly waiting for him to return as our Savior. (Philippians 3:17-21, NLT)

In a quote above it says,

> He also says to the Son, "In the beginning, Lord, you laid the foundation of the earth and made the heavens with your hands.

> "They will perish, but you remain forever. They will wear out like old clothing.

> "You will fold them up like a cloak and discard them like old clothing. But you are always the same; you will live forever."

I just thought it bore repeating.

Now that we know that Jesus is God (*the same God Who caused the Flood*), according to the Bible, we should think about what is going to happen when He comes back. And He has already told us what to expect.

When God Comes Back

Don't forget what you have learned already in this short presentation; God broke the planet because of man's depravity. Jesus is God. When Jesus comes back, He is going to punish the world for their treatment of each other and especially His followers. In fact, He has been quite clear about this for a very long time now. All of the Major Prophets discuss His return. And all of them talk about the pain of it for the planet. When Jesus comes back, there will be hell to pay—literally.

The possible physical destruction of Earth that I have warned about, through the melting under our feet, won't happen in your natural lifetime; but the return of Jesus very well might. And that is going to be very catastrophic for some!

This isn't doomsday nonsense. It's common knowledge among those who read Bible prophecy. Even people who are somewhat disconnected from the things of God feel as if something big is nearly here. And the Bible tells us what those things are.

As our ultimate Judge and Authority, Jesus will hold court for all of His human brethren (*He is Judge*). In the balance is their eternal destiny (*whether they like it or not*).

8 | The Promise

In another book, *End Times Made Easy*, I talk about God's judgment and how it is very misunderstood these days in the Church, much less understood in the darkness of those who are outside the kingdom of Christ. I'll share a couple of concepts.

Humanity 2.0

In a post on my blog, I talk about what will happen when Jesus returns. It's also in a previous book of mine, *End Times Made Easy*.

This is what Paul calls "the hope of our salvation"; it's that something we are all looking forward to as saints.

When Jesus comes back, He will transform all of His followers, whether dead or still alive, into super humans, like Him. We won't be equal to Him, but we will "share in His glory". It's like an upgrade from the manufacturer. That's why I call it **Humanity 2.0**. He will upgrade us from these dying, mortal bodies (*Humanity 1.0*), and give us bodies that are both physically and spiritually superior. Personally, I think that we will have powers of telepathy (*as Jesus uses today with us*), telekinesis (*tell this mountain to move*), and teleportation (*like what happened to Phillip with the Ethiopian, or Jesus entering the upper room that was locked shut*). We may also be able to soar (*as He did when He left the first time, going up to heaven and as He will do upon His glorious return*).

As a people, we will be one with God in a way that we don't yet understand. We will be invincible, as Jesus is invincible. We will know whatever we need to know when we need to know it. And we will know Him, as He knows us.

Jesus is Humanity 2.1. We will always be less than Him. Always. But just as the Father has lifted Jesus up to being One with Him, sharing in the same Holy Spirit, so too we, who are lower than the angels now, may be lifted up to share in Jesus' quality with His Holy Spirit in us both (*Him and each of us*) above the angels.

The Millennium

I also talk about this in *End Times Made Easy*. The millennium is a prophesied thousand-year period wherein humanity will be present in three different forms: Jesus (*the God-Man*); Saints (*The body of the Holy Spirit and the Bride of Christ*); and the Nations (*mortals who obey God and saints*).

Sheep to Nations
Green Pastures in a Perfect World

Jesus is God in a human body. He will always be above His creatures. There is no comparison between His perfection and the rest of us, with our imperfection. BUT, when we are glorified with the same glory that He has received from the Father, we will be perfect as well. (*This is not easy for me to imagine.*)

So in the millennium, Jesus will be on Earth (*Jerusalem*) sitting on His throne in His new temple (*different from all the others—much bigger, nicer*).

He will rule over the affairs of mankind on the planet, with His saints as His helpers. These are His faithful followers who share in His power, knowledge, wisdom, and grace. They will rule over the affairs of those people who are called "The Nations". They became that way through faith, not works.

The Nations are mortal people who never gave their lives to Jesus but didn't give their lives to Satan either. These are the ones who never took sides but never really followed anyone, either. They are called the "Sheep", from the separation of the sheep and goats that occurs at the beginning and end of the millennium (*two judgments*). They will be good citizens in Jesus' kingdom on Earth, but they are required to worship Him in specific ways. It's more of a religious, formal, legal relationship than being one with Him like the saints. It isn't much, but it is required of all nations to go before Him at least annually. If they don't, then they will not get any rain for their crops until they do.

Jews will be in the Nations. In fact, theirs will be the preeminent nation on Earth during the millennium. They will all live in Israel as given to them in the Bible. Every Jew on the planet will live in Israel. No exceptions. See Ezekiel, chapters 40 through 48. The Jewish religion lives on too in the Millennium.

People of the Nations will have families and produce offspring. Saints will not. Saints will be one with Jesus and His Holy Spirit. Perhaps we may even be one with the Father at some point. (*Maybe that's just semantics, since Jesus and Holy Spirit are God.*)

People in the Nations are mortal and can die on this Earth (*I don't know what happens when they die*), but they will not die in the New Earth; just in the millennium on this Earth. They are kept alive by drinking the Water of Life and eating the fruit from the Tree of Life (*the same one that Adam and Eve were kept from having*).

In a way, the Garden of Eden is returned to Earth in the form of the Holy of Holies in Jesus' temple. It's His throne room. Will Jesus perhaps remove the radiation as well?

This is a pre-taste of heaven on earth that we will have in the New Earth; the final home of all living mankind. But that will come after the Millennium is over and Jesus destroys this creation to build the new one. There will be tears and rebels in the Millennium, but not in the New Earth.

Life on Earth under Jesus will be an absolutely utopian existence. No wars will be able to be fought. All infractions of God's law will be dealt with using an iron rod (*absolute rule*). I don't think that anyone will be able to get away with any kind of crime, since the first responders and judicial officials will know every detail of the crime before it happens. There won't be any investigations where criminals get off free. No one will fool God and His government. You can't lie to someone who can read your mind or just see the whole thing occur in their mind or spirit. Resistance is futile. No pleads.

No deals. Just repentance and restoration; or destruction, I guess.

Judgment

Jesus' judgment of mankind will commence immediately upon His arrival to this doomed, broken rock—while still in the atmosphere, actually. The Great Tribulation will be wrapping up globally (*not in a good way*) and at just the last minute, when there is no hope left for humanity, it seems, there He will be, arriving from the East; riding on a white horse and sending His angel armies to both conquer and rescue. He will conquer the enemy and He will rescue His own who still cling to life and faith.

The rest of humanity—those who were caught in the middle—will be rounded up and stand trial to receive their sentence or reward. He will give to everyone as they have deserved.

> "Look, I am coming soon, bringing my reward with me to repay all people according to their deeds.
>
> "I am the Alpha and the Omega, the First and the Last, the Beginning and the End."
>
> "Blessed are those who wash their robes. They will be permitted to enter through the gates of the city and eat the fruit from the tree of life.

"Outside the city are the dogs—the sorcerers, the sexually immoral, the murderers, the idol worshipers, and all who love to live a lie.

"I, Jesus, have sent my angel to give you this message for the churches. I am both the source of David and the heir to his throne. I am the bright morning star."

The Spirit and the bride say, "Come." Let anyone who hears this say, "Come." Let anyone who is thirsty come. Let anyone who desires drink freely from the water of life.

And I solemnly declare to everyone who hears the words of prophecy written in this book: If anyone adds anything to what is written here, God will add to that person the plagues described in this book.

And if anyone removes any of the words from this book of prophecy, God will remove that person's share in the tree of life and in the holy city that are described in this book.

He who is the faithful witness to all these things says, "Yes, I am coming soon!" Amen! Come, Lord Jesus!

May the grace of the Lord Jesus be with God's holy people." (Revelation of Jesus Christ 22:12-21, NLT)

Let us not water down the message of our supreme God and Savior, Jesus the Messiah and Creator. If we take away mention of the pain, while talking only about the joy, we are not telling the whole story. It's the pain that makes the joy so joyous. Likewise, if we don't mention the joy, then the pain has no hope to pull us through.

What the majority of people in the world are getting from the world is no hope at all. That is because the removal of God, with His judgment, His invisibility, invincibility, and elusive presence, power, and rules, is the removal of any good at all. Meaning, there would be nothing left but pain, sorrow, regret, sadness...you know, the stuff that hell is made of.

Actual hell is an individual state, btw; no buddies to hang with in your misery. No drugs to comfort your mind in another state. No suicidal escape into another dimension. All you can do is wait for Him to judge you for all that you have thought, said, and done while walking the Earth. After that, it will only get worse for you when He judges you as you have judged others.

Did you get that big hint?

"Do not judge others, and you will not be judged.

For you will be treated as you treat others. The standard you use in judging is the standard by which you will be judged.

"And why worry about a speck in your friend's eye when you have a log in your own? How can you think of saying to your friend, 'Let me help you get rid of that speck in your eye,' when you can't see past the log in your own eye? Hypocrite! First get rid of the log in your own eye; then you will see well enough to deal with the speck in your friend's eye." (Matthew 7:1-5, NLT)

The destruction that Jesus wreaks upon His return is one kind of judgment (*while still in the air*); but then He will hold court and "separate the sheep from the goats" (*when He lands on the Mount of Olives after 5 months of wrath-giving in His aerial campaign of shock and awe*). In this judgment, He will decide who gets to live on Earth for another thousand years, basically.

These people are not the ones whom He calls His elect, or chosen, or saints, or followers, or disciples; instead, these "sheep" will be those people who are still alive on Earth at His Return and have not worshipped the antichrist or Satan. But neither have they followed Him. So they are a new category for us in the family:

The Nations| mortal people living on and on by not having radiation killing them any longer. And they are eating and drinking fruit and water that miraculously (*in our minds*) sustains their physical bodies indefinitely. (*Look up the Water of Life and Tree of Life*.) Adam and Eve were kept from it. These people will come alive with it.

"'In the last days,' God says, 'I will pour out my Spirit upon all people. Your sons and daughters will prophesy. Your young men will see visions, and your old men will dream dreams. In those days I will pour out my Spirit even on my servants—men and women alike— and they will prophesy. And I will cause wonders in the heavens above and signs on the earth below—blood and fire and clouds of smoke. The sun will become dark, and the moon will turn blood red before that great and glorious day of the LORD arrives. But everyone who calls on the name of the LORD will be saved.'" ~ Peter (Acts 2:17-21 NLT)

What I am doing with this presentation is the same thing that has been going on since the first century; I'm sharing what God has given to me with whomever will listen (*or read, or watch*).

As you just read one of the apostles announce publicly, God is working through His people in many ways—some of them miraculously. What I am doing here falls right in line with this teaching from the New Testament. God gave me an incredible vision in my mind of what the Earth looked like before the Flood, which He used to destroy those who had turned so far from Him that they would never come back. After that, He taught me what it all meant.

Some even think that the human race was corrupted with some kind of demonic half-breed situation, and that Noah was the only one left pure (*making him holy; set apart; different*). I don't know if that's true or not. But I do know that God ruined the world because of humanity's sinful ways. And He told us that in the time of His return we will be that evil again.

"When the Son of Man returns, it will be like it was in Noah's day. In those days before the Flood, the people were enjoying banquets and parties and weddings right up to the time Noah entered his boat. People didn't realize what was going to happen until the Flood came and swept them all away. That is the way it will be when the Son of Man comes." (Matthew 24:37-39, NLT)

They didn't even know that they were breaking God's heart with their ignoring Him and His wishes all the time. They were wicked and clueless about it. They were just as mean and nasty to each other as we are today.

But He let them have it. They were wiped from the face of the Earth in just minutes. All people on Earth (*likely in the millions or tens of millions*) except for one man and his three sons and their four wives, everyone just tossed into the mud that mostly hardened into rock. Death came quickly and mercifully; unlike the manner in which we killed Him on the cross (*slowly; torturously*).

Today when we receive that story, we react to it in various ways. Some of us scoff and or get offended and even make up other stories (*or gods*) to drown it out of our minds. Some of us are indifferent and shrug it off. Others listen, believe and want to change and makeup with God, whether out of fear, guilt, or repentance. Those are the ones He seeks.

I personally think that the Flood story was so fervently told to the younger ones by the older ones that it scared them (*as it was probably meant to*). I mean, what would have been on your mind to tell the young ones after living through the flood?

Then when Noah's extensive family was all split up by God at Babel and went their different ways, that's when different stories began to be told in the many lineages around the globe. Some kept it close to the truth; others veered away into some other story (*made up, of course*).

The languages were created by God to confuse the people. They got confused. And their whole way of life and society was torn to shreds overnight. Can you imagine the confusion, frustration and isolation of leaving the home, society, and family you loved because no one could communicate or get along?

You grab your metal axe and knife and whatever other tools and hardware you have, but you don't know how to make them yourself; you got them from someone else. You don't have all of the combined skills and knowledge of the society (*family, really*) that you are leaving. You and your closest family clan who can understand each other only know so much combined—not nearly as much as the whole family combined (*all of mankind at that point*). What a hit on technology! What a hit on society! What a hit on the truth!

Remember this; the human family immediately after the flood was probably very close-knit for the first few generations. Literally everyone you would have met was family, under a righteous patriarch. People tend to pull together in and after hardship. It was a really great society, I'm sure; it had to be the best on Earth so far and since, when it was all broken up. You know? Here everyone is getting along really well and enjoying life together for the first time, when God broke it all up. Man. That must have been heartbreaking and earth-shattering for most, if not all of them.

And so their response to God's heavy hand (*again*) was to pretend that He didn't even exist (*for many of them, anyway*). Some took up other gods (*demons, really*), that seemed to be more sympathetic to the wandering emigrants (*at first*). A few, I'm sure, stayed true to Yahweh (*not knowing His name yet*).

And so here we are; four thousand, three hundred and whatever years later, having to deal with our ancestors' stubbornness and refusal to follow God and His wishes. And the stories have just gotten more out of hand as time goes on. (*Could that be because man only becomes more corrupt as time goes by? Our DNA sure does. We can prove that.*)

So which story do we believe? Well, for me, not any that has Deep Time in it. In fact, the Bible has been so well proven to be true in everything that it discusses; I wouldn't ever toss it for another explanation at this point.

And I'm adamant about it (*zealous, even*) because this stuff has real world ramifications and consequences. There really is a God Who punishes sin and I don't want to be on His bad side or in His way when He comes back here. (*Actually, I'm looking forward to His return, as scary and horrific as it will be for others.*)

8 | The Promise

While Paul was waiting for them in Athens, he was deeply troubled by all the idols he saw everywhere in the city.

He went to the synagogue to reason with the Jews and the God-fearing Gentiles, and he spoke daily in the public square to all who happened to be there.

He also had a debate with some of the Epicurean and Stoic philosophers. When he told them about Jesus and his resurrection, they said, "What's this babbler trying to say with these strange ideas he's picked up?" Others said, "He seems to be preaching about some foreign gods."

Then they took him to the high council of the city. "Come and tell us about this new teaching," they said.

"You are saying some rather strange things, and we want to know what it's all about."

(It should be explained that all the Athenians as well as the foreigners in Athens seemed to spend all their time discussing the latest ideas.)

So Paul, standing before the council, addressed them as follows: "Men of Athens, I notice that you are very religious in every way, for as I was walking along I saw your many shrines. And one of your altars had this inscription on it: 'To an Unknown God.' This God, whom you worship without knowing, is the one I'm telling you about.

"He is the God who made the world and everything in it. Since he is Lord of heaven and earth, he doesn't live in man-made temples, and human hands can't serve his needs—for he has no needs. He himself gives life and breath to everything, and he satisfies every need.

"From one man he created all the nations throughout the whole earth. He decided beforehand when they should rise and fall, and he determined their boundaries.

"His purpose was for the nations to seek after God and perhaps feel their way toward him and find him—though he is not far from any one of us. For in him we live and move and exist. As some of your own poets have said, 'We are his offspring.'

"And since this is true, we shouldn't think of God as an idol designed by craftsmen from gold or silver or stone.

"God overlooked people's ignorance about these things in earlier times, but now he commands everyone everywhere to repent of their sins and turn to him. For he has set a day for judging the world with justice by the man he has appointed, and he proved to everyone who this is by raising him from the dead."

When they heard Paul speak about the resurrection of the dead, some laughed in contempt, but others said, "We want to hear more about this later."

That ended Paul's discussion with them, but some joined him and became believers. Among them were Dionysius, a member of the council, a woman named Damaris, and others with them. (Acts 17:16-34, NLT)

That scene has been played out in many college campuses and social clubs around the world. A man shows up, talks about a new philosophy and people follow him or his view, method, or concept, while others don't. Some of those views have been very different.

Today, the average human is bombarded by more philosophies and concepts and theories than he can bear, *present activity included*. But how does he tell the truth from the fiction? The enemy has torn up the family as much as he can, and along with it a functional society. The things he throws at our youth are just incomprehensible sometimes.

Today, our children possess very weak critical thinking skills, connection to family, or a larger society to feel safe in, or a connection to their Creator, thanks to the schools.

All of these things have been stripped from our children at school, while we both parents go off and live independent lives of each other in some corporate, business, or scholastic environment all day (*less, now with telecommuting gone wild*).

The attack on our kids has been so severe that they have the kids doubting what a gender is; *of all the certainties in life*. If they can doubt such an obvious thing, then the higher and already barely attainable truths are going to just fly away from them, always beyond reach or even memory. Whatever you may think of the enemy, his diligence is paying off for him—to our detriment.

The enemy is in school more than anywhere else in America. As a ruler in the world, Satan has given a certain kind of authority to the schools that has never really been manifest as it is today. If a school gives you a certificate that says "PhD" on it, you can make more money than those who don't (*unless your name is Bill Gates, or Elon Musk, then you just forego the PhD altogether.*) But in the process, you become indoctrinated into his lies.

Why are we not reeling-in the money flowing to those schools that are out of control? Why are we not governing the schools, their message, and behavior? Why are they getting a free pass to ruin our children's minds?

I'll bet that if I were allowed to follow any student's schedule through any public school in the US, I'd see more outrageous behavior than I would be able to bear. And that from just the faculty and administration. If I were there for a week, I'd probably want to shut the place down (*if I survived*).

Conclusion

I'm not a scientist. I'm not a prophet. I'm not even that smart compared to many of you. But I understand

what I believe God has shown me here. Using this model, I can see how the Earth got to be the way that it is now. In studying the Bible, I've learned a lot about Cosmology *(study of the beginning)* and Eschatology *(study of the end)*. In studying the Earth, I see God's fingerprints all over it.

I can now say with confidence just what the story of Earth was and is and is going to be, because my Best Friend showed me a little more than what He put into print (*the Bible*). He was there and knows it all. This little bit that He shared with me about cosmology (*a 5-to-8-second vision with follow-up insights*) has been mind-blowing for me, personally, and I thank and praise Him for His knowledge, grace, and love that He has shared with little, insignificant me.

I think that books could be written about God's reasons for breaking the planet like and when He did. If nothing else, it sure makes a statement. What that statement is, exactly, seems to depend upon the one receiving it.

The mystery is solved. We can now see how it all came about and where it's going. We can quit spending money on figuring out the origin of life and instead help orphans and widows who are already here and in need. With this model, we can see that the world could have been quickly created in an amazing design centered on our comfort. We can see that James Hutton was wrong; the past **has** determined our present and sealed our future as well.

This Broken Planet

The Earth is young—and it will die young. To deny this is to deny the evidence beneath our feet and in the testimony of God's Bible from the One Who made it all.

We've been very patient with Mr. Hutton and his disciples for a very long time now, but I think it's time to talk about something else in school when it comes to Cosmology. This *Deep Time* nonsense has gotten old—and rotten. And while we're at it, why not remove the cult as well? (*Those satanic administrators who hate God and His kingdom.*)

> Yet what we suffer now is nothing compared to the glory He will reveal to us later. For all creation is waiting eagerly for that future day when God will reveal who His children really are. Against its will, all creation was subjected to God's curse. But with eager hope, the creation looks forward to the day when it will join God's children in glorious freedom from death and decay. For we know that all creation has been groaning as in the pains of childbirth right up to the present time. And we believers also groan, even though we have the Holy Spirit within us as a foretaste of future glory, for we long for our bodies to be released from sin and suffering. We, too, wait with eager hope for the day when God will give us our full rights as His adopted children, including the new bodies He has promised us. We were given this hope when we were saved. (If we already have something, we don't need to hope for it. But if we look forward to something we don't yet have, we must wait patiently and confidently.) (Romans 8:18-25, NLT)

8 | The Promise

I have written this book (*or post*) for you so that you will consider that science doesn't remove God from the picture; you do, when you allow bad concepts (*that remove God*) to rule your mind and harden your heart against Him.

If you spend your life ignoring God, then He'll let you do that for all eternity after you die. Initially, it's called "hell", after that is the Judgment of everything you've ever thought, felt, said and done on Earth. Then it might be the Lake of Fire for you, which is never going to go away.

If that is your destiny, then this world is the closest thing to heaven you're ever going to experience. On the other hand, if you become a believer (*before you die*), then this world is the closest thing to hell that you will ever experience.

It really is up to you. Give in to Him, or not. He's waiting for your answer. And it's not like you've got a million years to mull it over.

In Him, and for you, not against you,

Gary

Group Discussion Questions:

1. Why did God ruin His beautiful planet with an explosion and Flood?
2. Did God just leave us on our own, or has He been with us all along?
3. When will Jesus come back to us?
4. Is the offer still open for you to enter in to eternal happiness through His grace?

Answers:

1. Because of the evil that mankind was living out from day to day in ignorance of His love for them.
2. He has always been with us, waiting for us to come back to Him for loving fellowship.
3. At about the 6,000th year of Earth's existence, which is not very far away now.
4. YES! This very day you can enter into a wonderful relationship with Him through Jesus, His loving Son, our Savior. All you need to do is believe in Him and turn from your evil inner depravity.

Chapter 9 | Model Summary & Q&A

Before jumping into the Q&A, I want to give a very concise SUMMARY of the Broken Planet Model. Hopefully, this will aid the reader as a quick reference guide to the main points of this new way of seeing Geophysics.

Core Premise

The Broken Planet Model (BPM) presents a catastrophic **one-time, inimitable (not repeatable) global Flood event** that reshaped the Earth, aligning perfectly with biblical history while dismantling deep-time assumptions. The model explains **geology, meteorology, hydrology, plate movement, fossilization, radiometric dating, the Ice Age, and continental formation** as consequences of this singular, inimitable event.

1. Pre-Flood Earth

- **Single, continuous landmass**—no separate continents, just one vast, lush, green land with some surficial water features.

- **Pre-Flood terrain was much flatter**—no towering mountains or plunging canyons, only shallow hills, valleys, and wrinkles.
- **All water was fresh**—no pre-Flood saltwater; salination happened post-Flood as minerals leached into the post-flood water.
- **Pre-Flood seas possibly extended to bedrock**, allowing direct water movement between the surface and subterranean layers.
- **The Moon may have influenced the water layer**, causing periodic water rise and fall via gravitational pull.
- **Pre-Flood Earth was a perfect sphere**—no equatorial bulge because the mantle was intact and firm.

2. The Pre-Flood Water Layer

- A **fully liquid subterranean water layer** existed **between the bedrock and the mantle**, estimated at **1 to 2 miles deep** globally.
- **Extreme pressure** existed, as the water was contained, and it was **superheated** due to the heat of the mantle.
- Water transferred through **porous bedrock and/or vents**, replenishing surficial lakes, rivers, and seas.
- The **Coriolis effect kept the water in constant motion**, similar to Jupiter's atmospheric bands.
- **No life existed in the subterranean water layer**—the heat. lack of oxygen, and turbulence would have prevented survival.

9 | Model Summary & Q&A

3. The Impact Event: Trigger for the Flood

- A **large, dense celestial object (meteorite)** struck Earth near **modern Iceland, punching through the crust into the mantle**.
- The **impact was nearly perpendicular**, creating a **deep penetration rather than a surface crater** (see Icelandic Plume).
- A crater would have been impossible since the crust it hit was almost immediately demolished.
- **Iceland and the Icelandic plume** are the surviving remnants of the impact site.
- **The collision did not shatter the crust outright**—instead, the release of **lava and water interaction** caused an **explosive chain reaction** within it.

4. The Global Crust Collapse & Flood Mechanics

- The **mantle was breached**, and **superheated lava met the subterranean water layer**, triggering **massive steam explosions** in a sealed chamber.
- The **explosions spread globally in minutes**, causing the **entire crust to bulge, then fracture and collapse into the water layer**.
- **Walls of water erupted from the cracks**, forming **towering water jets that fell as rain**.
- The **collapsed bedrock plates sank onto or into the mantle**, while **lava oozed between them**, forming **new landmasses** atop a new Asthenosphere.
- **Seafloor spreading today is just residual lava leakage**, not plate tectonics as conventionally taught.

5. The Immediate Post-Flood World

- The **world transformed from a lush green land to a muddy, turbulent wasteland** in mere days.
    - Green Planet—Lush, continuous biomes globally
    - White Planet—Continuous cover of towering storm clouds
    - Brown Planet—Mostly covered by mud
    - Blue Planet—What we have today, mostly water
- **Vast waves and cycles of water motion** reshaped the planet, layering mud over new landforms founded on lava.
- **Continental landmasses formed in situ** as **mud hardened over lava flows**.
- **Fossils and fossil fuels** resulted from rapid burial in mud, which quickly baked into rock.
- **Massive volcanic activity** persisted, blocking sunlight and triggering global cooling.

6. The Ice Age: A Direct Result of the Flood

- The **atmosphere became supersaturated with water vapor**, causing **instantaneous extreme precipitation**.
- **Poles froze rapidly**, down to about the **40th parallels**, north and south.
- **The Ice Age was a singular, inimitable event, not a repeating cycle**—it happened due to the unique post-Flood climate conditions.
- **"Glacial scars" are actually flood scars**—not caused by slow-moving ice but by the final flows of water draining into ocean basins.

9 | Model Summary & Q&A

7. Post-Flood Human and Animal Migration

- **The Ark landed in the mountains of Ararat,** below the highest peaks on the southern slope.
- **Humanity remained together in Mesopotamia** until the **Tower of Babel incident.**
- After Babel, **rapid global migration** occurred—civilizations emerged **within Noah's lifetime,** worldwide.
- **Central American pyramids/ziggurats were likely built by early post-Babel migrants** while Noah lived.
- **Animals spread based on their environmental tolerances,** not evolutionary adaptation.

8. Radiometric Dating: The Great Reset

- **A massive inimitable radiation event occurred during the Flood,** when **radioactive elements were released from the nuclear mantle** and outer core.
- **Before this event, Earth had almost no measurable radiation**—modern radioactive decay rates (entropy) are just the **residual effects of this catastrophe.**
- **Radiometric dating assumes constant decay over deep time,** but this model (BPM) shows that **radiation was dumped in one single event.**
- **Just like fossilization, the Ice Age, and crustal collapse, radiometric dating is based on a false backstory.**
- This explains the **rapid decline** in human life expectancy, from over 900 years to 120 years, max.

9. Marine Survival Hypothesis

- **Marine life survived in a few or single, large pre-Flood seas** (or possibly several seas), which remained **relatively uncontaminated**.
- **Survival depended on location**—species in high-mud, high-heat zones perished, while those in stable pockets survived.
- **God could have divinely protected the refuge pockets** to ensure the preservation of necessary species.
- **Since all pre-Flood water was fresh**, marine life had to **adapt to salination post-Flood** as ocean chemistry changed.

10. Magnetosphere Disturbance

- The mantle rupture and impact would have **destabilized Earth's magnetic field** due to material being sent into the outer core, disrupting the dynamo's flow.
- The resulting disturbances could explain
    - **magnetic pole shifts**
    - **variations between true north and magnetic north**
    - and **imprinted magnetic anomalies in surficial rock layers.**

11. Cosmic Debris Hypothesis

- Some **comets** may be comprised of debris ejected from Earth's explosion, possibly launched into space by the Flood catastrophe.

9 | Model Summary & Q&A

- A possible **debris field** may have existed in Earth's orbital path, potentially responsible for later meteor showers and impact events.
- Or, it may have impacted the Earth in subsequent orbits through the field.

12. Evidence Emerging Today

- **Modern mantle anomalies** (cold zones, debris-like structures) **support this model**, as they could be remnants of **mantle collapse during the Flood** due to the chain of explosions that ruined it.
- **The Mid-Atlantic Ridge and other rift zones are still oozing lava**, proving the crust is fractured and still adjusting.
- **Deep-sea trenches, volcanic islands, and oceanic ridges** align with the catastrophic rupture event.

Final Thoughts

The Broken Planet Model systematically **dismantles deep-time assumptions** by providing a **cohesive, one-time catastrophic framework** that elegantly explains Earth's features far better than mainstream science. It is internally consistent, agrees with scientific data, and biblical revelation like no other model.

Questions & Answers

Where does radiation come from?

According to my Broken Planet Model (*BPM*), it should all be down in the Mantle, safely locked away, but the Mantle broke, letting it come out. We even know the exact date that this world was irradiated by itself: It came with the Flood of Noah, which began on the 17th day of the 2nd month in 1,656 *anno mundi*. The Sun is a source of radiation as well. Is there cosmic radiation? idk

Doesn't radiometric dating disprove the Bible's timeline?

No. Radiometric dating cannot measure the initial amounts of sample isotopes in the past. Since the "process" depends on the difference between counts, it is useless (*there is no process; just a little man behind the curtain, named Al Gore-Rhythm*).

The main problem with this method is that the ones measuring the isotopes don't know that the overwhelming amount of radiation we have came upon the planet very rapidly. about 4,300 years ago. They believe that it came on very gradually from solar and cosmic radiation. That is the main problem with it.

Carbon dating is also a guess at best, being unable to account for outside influences on the sample either, thereby being wildly inaccurate, for Great Flood victims especially.

9 | Model Summary & Q&A

Where was the water before the flood?

We don't know. This BPM—Broken Planet Model— is my explanation of what God showed me—a layer of water between the Mantle and the Crust. When the Crust broke, the Water Layer spilled out—like when breaking an egg. At the same time, causing its breakage actually, the Mantle was broken with its "skin" being blasted into its inner parts (*at least in places*), thereby exposing the shattered Crust to the heat and radiation below.

If the Mantle had not broken, exposing super-hot matter from within, the soil from the broken Crust would not have gotten baked into position as it is, building up the continents above the water line. That turned out to be an important feature for our survival. If the mud had not baked, it would not have hardened, and the land would not have gotten built up enough to emerge from the waters. Huh, I just realized that the Potter (*a reference to God*) baked the continents in place so we could live. We could say that the Potter mixed lava and mud together to make us a new home.

And since the land arose with the mud and lava, the water had a place to recede to (*coming off of the rising land*). No heat equals no hardening and building up of the mud. Without this hardening, the mud would just be a soft mud layer under a water world. But God is merciful to us rebellious humans, even in His most vengeful wrath against our impertinence toward Him. He could have just drowned us all. And if the Crust Layer broke but the Mantle did not, that's what would have happened. But because the Mantle broke along with the Crust, and the heat did its thing…

Why are there fossils?

There are fossils because all lifeforms (*flora and fauna*) were swallowed up by the ground that was turning to mud and getting baked by the lava. The heat from that lava emerging from below, along with the pressure of the mud's weight above, baked impressions of the lifeforms into the mud as it hardened into stone. In some instances it seems as though the lifeform itself was turned to rock. *Just like Lot's wife, I guess, but different, I guess.* We find both positive and negative impressions of plants and animals in the rocks under our feet today. But my guess is that Earth's pre-flood life was mostly turned to oil or coal or flammable gas. Well, and nutrients in the soil too, if close to the top, I suppose.

What was Earth like before the ruin?

Perfectly suited to support life indefinitely. Humid. Warm. Always gentle weather. There is no way of knowing what the surface features were, since the entire surface was churned into a different shape during the breakup of the Crust Layer. Based on the amount of soil now above the broken Tectonic Plates (*of the former Bedrock Layer*), I would just guess that the Soil Layer was rather thin compared to our continents today (*if it were spread out all around the larger globe, especially*).

9 | Model Summary & Q&A

Why are there ethnicities?

We see ethnicities because the genetic lineages started at the Tower of Babel, when Noah's family was broken up, and family genetic lines grew shallower from more isolation or less diverse contact from then on between the emerging, diverging clans.

Why are there different languages?

God created them because Noah and his family were not spreading out to cover the Earth like God had told them to. I say Noah, because according to his lineage in the Bible he was still very much alive at the time of the Babel incident. Why would he not be with his family? I would expect Noah to be the world leader in some fashion or other. He was the oldest and probably wisest of the entire human family (actually, idk how old his wife was). He was the last of the true "ancients" in our lineage; THE Patriarch of the day, charged with the stewardship of Earth. He was there (freaking out when the *languages hit, no doubt*).

Why are there so many religions? And why are they so similar?

They aren't all that similar; just from similar sources (*men and demons*—except one, which is from God). When Noah's family was split up abruptly, and in a very scary fashion (*supernaturally imposed*), on the heels of the Great Flood (*by 101 years*), it may have been too much for some people to take. Many probably developed some kind of animosity toward the Creator. Or perhaps the truth just got eroded in time, along with the crops and villages. The common thread in the tapestry of most religion is this: *Human Effort*. Judaism, arguably the only religion started by God, was presented as an alternative to a faith-based relationship with God (*which He led with in Abram*). Christianity is a continuance of the relationship that has been born from Abraham; it is not religion (*if you think it is or should be religion, read Galatians*). Nearly all religious systems demand a certain level of commitment and servitude from its adherents. Since the earliest settlers of the cradles of humanity were all related (*first cousins*), it would not be unusual for many common ideas to exist in almost all early religions around the world. I believe this, because they all spread out so far and quickly from Babel. All people groups on earth can trace their family origins back to the people listed in Genesis chapter 10, the Table of Nations. These are the cousins of Peleg, who were born when the language division hit.

Also, at that time, the family was very preoccupied with fame and accomplishment, as in with Nimrod and/or building the great tower (*ziggurat or pyramid*). The building of something monumental was going to be a great achievement in their eyes. God did not mention removing their ambitions; just splitting them apart by confusing their speech *(too bad they didn't keep a married couple from each language there together to learn the languages from—a world language institute, if you will—keeping their languages alive with their clans nearby that could go off to where the others went…)*.

9 | Model Summary & Q&A

Anyway, we can see monumental structures that all went up very close to the same time all around the world in the most ancient cradles of society. I would be interested to see this shown through further Y chromosome collection and assessment; not the other failed dating methods. And give it to Dr. Jeanson for his research.

The neo-religious (*some would say "non-religious"*) aspects of both Deep Time (*millions of years*) and Evil-You-Shun (*evolution theory*) are exceptions to the older religions and faith systems with their rules and demands. Or perhaps they both are two aspects of the same mindset (*humanism, for example*). These theories, turned philosophies, turned beliefs, turned paradigms (*which are one in my mind in this light*), turned cult, have been wildly attractive to people all over this time and space continuum because these new belief systems allow any lifestyle for its adherents to follow (*or not*).

This cloudy mindset completely removes God and His rules for our behavior from our lives and minds and hearts (*like covering one's eyes to make the monster disappear, I guess*). We think that we are free from tyrannical religious law when we subscribe to such philosophies. However, the truth is that such a mindset is but a mental and spiritual trap for its victims, since it denies reality that exists on, beneath, and far above the surface (*"to infinity, and beyond"*).

And a detachment from reality is not a good state to be in. Who knew that our entire society (*quite nearly*) is living in a dream world, detached from reality? This ends up being a means of separation from the Creator. And I'd welcome detachment from reality if it led to unity with Him, but I'm not sure that's the most favored path to take. Either way, sane or not, detachment from God is to be avoided at all costs, including one's sanity, and life, and limbs, if it should come to that. But before you cut something off, remember that nothing can come between Him and His. Just make sure that you're one of His, so that He doesn't cut you off altogether.

On Religion

> There is a way that seems right to a man, but its end is the way of death.
> (Proverbs 14:12, Berean Standard Bible)

"Oh no, I don't need your help God; I've got this. Let me show you."

This is not the way that Christ set out for us to follow. We are to follow Him and His efforts on our behalf if we are to make it to the Father in good standing. Trying to get there on our own steam is futile, especially when we need to go through Jesus anyway, no matter what. He is both our Prosecutor and Defender. And He's really good at His job; either way it goes for us. And He is our Shepherd; we are the sheep. Sheep do not shepherd themselves.

"Divide and Conquer" is an old military axiom. For the common man, it addresses the fact that people are not as strong as individuals as we can be when united. God addressed that point at the Babel incident. This is how armies can take

9 | Model Summary & Q&A

over entire nations that outnumber them. They divide them into smaller groups that are easier to defeat.

Well, this is what our enemy took advantage of when God sent us in different directions from the great division at Babel. The enemy didn't scatter us, but he sure took advantage of the situation before him. People began to listen to the enemy as they did before the flood.

Demons became gods to them and they followed them and their ways. My own opinion on this is that almost all religions are demonic in origin. The only exception I can think of is Judaism, which was instituted by God, directly. But He did it to make a point— that it doesn't work. When Christianity becomes a work and reward system, it is in the realm of demonic theology, and should be shunned as such. In my opinion, turning God's gift of salvation into something that must be earned is the worst kind of evil there could be (*see Galatians 3:1-14*). Even worse than the horrors that would soon follow from such a condition (*see Church History*).

But the big takeaway here is that taking God's rules out of the picture also takes away His blessings—in this world and the next. The cost is far too high, especially since all things that are contrary to His ways are empty and not fulfilling at all, no matter what—and doomed to destruction.

Why do people avoid Christianity and Judaism, if both are supposed to be genuine? And why don't Christians and Jews agree?

Well, it's kind of the same question in my mind. The disagreement comes from believing or not believing the New Testament and specifically Who Jesus is. Those who are Jewish remain Jewish when they accept the Lordship and sacrifice of Yeshua HaMashiach (*Jesus*). Their Judaism doesn't go away, they just acknowledge that the atonement is taken care of for them now and they follow the apostles' teachings (*the Apostles were Jews, too*). Their Judaism doesn't save them from God's judgment. If you approach God with the Law (*works*) you will be judged according to the Law and your own judgments upon others; if you approach God with a contrite heart and faith in Jesus' works and not your own, then you will be received on your faith (*if it is genuine*). Faith leads to sainthood apart from judgment; I speculate that judgment by works at best leads to membership in the Nations on the New Earth (*if you pass*). As a member of "the Nations", you would be under God but not one with Him. He will hear and answer your prayers and be good to you as long as you obey Him and His saints. You would not share in His higher commutable abilities, though, like the saints will.

Becoming a member of the Nations, through works, seems much more difficult than becoming a saint on faith alone. And I do not consider attaining citizenship in the Nations as a goal;

9 | Model Summary & Q&A

it's more of a safety net. Like a trapeze artist, I want to be caught, mid-air, and not hit the net.

The reason why Christianity is not winsome to others is the behavior of the Christians is not attractive. I can't speak for the Jews but suspect the same condition to exist in their ranks as well; haughtiness can be an issue. So can obvious sin.

Why does Earth look old, eroded, and broken?

Earth looks old because radiation is breaking everything down at the molecular level. It is called "entropy" by the scientists. Erosion from the Flood also makes the surface look old. Breakage is from the Flood. Radiation contamination is from the flood. Jagged, rugged mountains in their purple majesty are from the flood (they all have seashells on their peaks, you *know*).

How could a 40-day rainfall flood the world, covering the highest mountains?

I couldn't, that would be silly. There needs to be a model like mine that has a large amount of water tucked away someplace out of sight, like the Bible says, which is suddenly released. Of course, my model also has the mountains being formed under the flood and emerging up out of it, or soon after.

As a quick note; the mountains where I live are evidence of the earth having been broken, with large pieces being pushed up from jostling, during the flood's turbulence. The layers' baking presented large masses of semi-firm rock being jostled about by the heaving waters, mud, or lava. They were soft and pliable, then became hard and brittle, and being further baked in place. A cooked layer is sitting in place where it was baked and then is suddenly broken and shoved to and fro, making all kinds of shapes in the forming new crust layer, like the Grand Canyon, for example. This also aids in piling the continents higher and faster into mountain ranges, plains, and plateaus. Lava flows add to the rock layers of the continents. Lava shoots up in the oceans to become islands.

Imagine baking cookies by placing big globs of dough on the hot baking sheet in random shapes and then continuing to mess with it (*stirring it*) the whole time it hardens and adding more and more dough as it sets; that might be like the formation of our sedimentary continental rocks in layers. Only, they were formed in water, partly, and tossed around by it for months. During and after the tossing, much volcanic activity took place as well, whether the water had receded by then or not. It was the combination of lava and the mud that built up the continents.

Some stratification shows evidence of soft and warm, not yet solid mud/rock being bent into wavy shapes. That is evidence of heat and motion working together on baking and hardening mud as it turns to rock. And we see fossils in this type of formation, too, like nuts & chocolate chips. Geological flood evidence like this is evident everywhere, evidently.

During the Flood, at the crest of the water level the land/mud elevation was 22′ less than the water level. But that was in flux and soon changed as the mud became dry and swollen, as did the lava.

How could one man put enough animals on a barge/boat/box to give us the species we have today?

He only needed "kinds" of animals (*similar to "families"*), which can lead to many species. Of all the kinds of air-breathing animals we now have, plus the ones extinct, including dinosaurs, there would have been enough room for them and all of their food for the year of the flood. It was big. And not just the food for the trip; but they will need to start all over in their gardening. And there won't be anyone else to get seeds or plants from. The old nursery that Noah shopped at would soon be at the bottom of the continent, so he needed to take everything he needed to start a new garden.

Noah wasn't alone. He could have assigned one son and wife couple to each of the three decks. Then he and his wife could roam around between the three decks, helping as needed (*for example*). It would have been a lot of work for them when things settled down enough to move about the cabin, but not overwhelming. I imagine that when they came to rest is when they had to get to work, as taking care of the animals ramped up.

What happened to all of the dinosaurs?

Well, there are actually many things to say about that. It could be a book on its own, really. But I can keep it short for a quick overview.

"Dragons" is what they were called in olden times by many cultures. The word "dinosaur" is pretty new (*ever since scientists started making stuff up about them, like, "they're mythical"*). And so it isn't so difficult to see that many of them were killed off by ambitious young men with weapons (*or men protecting their families and clans*).

They were probably not very tame animals. Reportedly, one in Congo several years ago had a nasty disposition toward people. And that was one of the "veggie-saurs"; a brontosaur. Imagine how thrilling it must have been to go after a T-rex or a Raptor as a young man or with your hunting buddies.

The Bible mentions two dinosaur- or dragon-type creatures in the book of Job in the Old Testament. Apparently, Job knew what they were when God mentioned them.

Leviathan

Leviathan sounds like a fire-breathing sea dragon. Fire-breathing is not unheard of in the animal kingdom. Mixing of two chemicals that are in glands in the throat or other area of an animal can facilitate that ability (*see bombardier beetles*). It wouldn't even have to be actual flames to be called "fire breathing"; the fluid could just be flaming hot or acidic to earn

9 | Model Summary & Q&A

the title. But actual flames could be made by a creature if that is God's design. He has that ability. He's that smart.

I'm going to share with you the entire chapter that discusses this dinosaur in God's talk with Job. As you read it, imagine an actual fire-breathing amphibious dragon, just like in the best movies (*with feet, not flippers*). You'll get the most accurate image that way. (*I quit reading the NIV regularly when they called it an alligator, and behemoth a hippo or elephant.*) We join God questioning Job.

> Can you catch Leviathan with a hook
>   or put a noose around its jaw?
> Can you tie it with a rope through the nose
>   or pierce its jaw with a spike?
> Will it beg you for mercy
>   or implore you for pity?
> Will it agree to work for you,
>   to be your slave for life?
> Can you make it a pet like a bird,
>   or give it to your little girls to play with?
> Will merchants try to buy it
>   to sell it in their shops?
> Will its hide be hurt by spears
>   or its head by a harpoon?
> If you lay a hand on it,
>   you will certainly remember the battle that follows.
>   You won't try that again!
> No, **it is useless to try to capture it**.
>   The hunter who attempts it will be knocked down.
> And since no one dares to disturb it,

who then can stand up to me?
Who has given me anything that I need to pay back?
Everything under heaven is mine.

"I want to emphasize Leviathan's limbs
and its enormous strength and graceful form.
Who can strip off its hide,
and who can penetrate its double layer of armor?
Who could pry open its jaws?
For its teeth are terrible!
The scales on its back are like rows of shields
tightly sealed together.
They are so close together
that no air can get between them.
Each scale sticks tight to the next.
They interlock and cannot be penetrated.

"When it sneezes, it flashes light!
Its eyes are like the red of dawn.
Lightning leaps from its mouth;
flames of fire flash out.
Smoke streams from its nostrils
like steam from a pot heated over burning rushes.
Its breath would kindle coals,
for flames shoot from its mouth.

"The tremendous strength in Leviathan's neck
strikes terror wherever it goes.
Its flesh is hard and firm
and cannot be penetrated.
Its heart is hard as rock,
hard as a millstone.

9 | Model Summary & Q&A

> When it rises, the mighty are afraid,
>   gripped by terror.
> No sword can stop it,
>   no spear, dart, or javelin.
> Iron is nothing but straw to that creature,
>   and bronze is like rotten wood.
> Arrows cannot make it flee.
>   Stones shot from a sling are like bits of grass.
> Clubs are like a blade of grass,
>   and it laughs at the swish of javelins.
> Its belly is covered with scales as sharp as glass.
>   It plows up the ground as it drags through the mud.
>
> "Leviathan makes the water boil with its commotion.
>   It stirs the depths like a pot of ointment.
> The water glistens in its wake,
>   making the sea look white.
> Nothing on earth is its equal,
>   no other creature so fearless.
> Of all the creatures, it is the proudest.
>   It is the king of beasts."
> (Job, chapter 41, NLT)

And Noah could have had a couple of young ones on his boat, if they had legs and feet instead of fins! How exciting would that have been to see? (*Feet might be required to be walked by little girls.*)

This Broken Planet

"Watch the hay around those fire-breathing dragons, Ham! We'll all be on fire if they sneeze!—and keep 'em happy and well fed, too!"

(Now I want a painting of a young pair sleeping together on the hay in the Ark.)

Behemoth

The other one, Behemoth, sounds like a kind of brontosaur, but a lot of modern Bible commentaries try to make it out to be a hippo or an elephant (*again, NIV comments fell far short*). I guess they didn't read their own translations. This is no animal with a tiny tail, like the tails of hippopotamuses and elephants. Here's what Job says about that creature.

> "Take a look at Behemoth,
>   which I made, just as I made you.
>   It eats grass like an ox.
> See its powerful loins
>   and the muscles of its belly.
> Its tail is as strong as a cedar.
>   The sinews of its thighs are knit tightly together.
> Its bones are tubes of bronze.
>   Its limbs are bars of iron.
> It is a prime example of God's handiwork,
>   and only its Creator can threaten it.
> The mountains offer it their best food,
>   where all the wild animals play.
> It lies under the lotus plants,
>   hidden by the reeds in the marsh.

9 | Model Summary & Q&A

> The lotus plants give it shade
>   among the willows beside the stream.
> It is not disturbed by the raging river,
>   not concerned when the swelling Jordan rushes around it.
> No one can catch it off guard
>   or put a ring in its nose and lead it away.
> (Job, chapter 40, NLT)

After the flood, there would not have been as much earth for men and dinos to keep away from each other on as there was before the flood (*from almost 100% to far less than 29%*). I'm sure that before the flood the people just stayed away from the potentially nasty-tempered giant reptiles. And so with less room for us to cohabitate, we ended up killing them off, because that's how we are. Of course, the weather was pretty bad for quite a while there too, after the flood, which might have affected their numbers. *Some dinos are encased in ice, while crossing a river—this means they lived after the Flood.*

There are ancient drawings of dinosaurs on rocks in different places in the world. There are tracks of dinos and people together in Texas. A plesiosaur, like the "Loch Ness Monster", was washed up on Monterey Bay in California in the 1930's, I think it was. A photographer managed to get some shots of it before the tide took it back out to sea.

I look at gators and crocks as tiny dinosaurs that are still with us (*not as much of a challenge or threat as the others, I guess*).

Soft tissue, called collagen, exists in many fossilized remains of dinosaurs. This is becoming more common as they dig up more fossils. But this should be impossible, if the tissue is to be 85 million years old. Soft tissue like that only lasts a short time; thousands of years, at the most. Oops. Well, it just isn't that old. The whole story has come unraveled for our college community. They are going to have to either give up their fantasies or admit that they are just blindly following a failed notion out of an emotional response to God and His Book.

Why is there still ice on the poles if it's all melting?

Because it's only been melting for about 4,300 years; not millions or even hundreds longer. When it's gone, it's gone. We won't have another Antarctica that is covered in snow and ice again. No more ice ages when this one peters out. Penguins might have to move to Mars. "Earth Penguins Invade Mars!" But Jesus will be back before Greenland is green.

9 | Model Summary & Q&A

How could a crust layer be suspended on a water layer?

(*When you zoom in on an image of layers, it's easy to forget that it's spherical.*) So, why doesn't the bedrock fall through the water layer? Well, how does a hollow ball hold water? The Crust doesn't fall through because it's firm and holds together since it is in a shape that naturally holds its own weight. The spherical shape of it counteracts the gravity pulling it down uniformly. The water has density and helps to hold it in place. The pulling force is uniform from gravity, not allowing it to go closer to the Mantle on one side or the other.

Earth's crust bulges at the equator now because it isn't held together anymore like when it was created. Its structure has been compromised and so the squishy parts are bulging at the middle for lack of support. I mean, the Mantle and the Crust were blown up! The Mantle went inward and the Crust collapsed upon what was left of it. No wonder it's bulging out at the middle. The explosion hitting the soft center did it.

Wouldn't the Water Layer boil and cause heat buildup and pressure?

I think it would to a degree (*pun*). I expect that vents were present to off-vent steam or pressure. The Bible mentions springs of water coming up from below (*See Genesis 2:6*). Noah used tar on the boat to seal it. Where did the tar come from? Hot steam vents.

Don't we have records that go back more than 6,000 years?

No. We have records, but interpreting them is not as straightforward as we might like it to be. And the presence of readable records does not ensure their accuracy. A more accurate technique is available to us today through DNA, whereby the scientist can measure the degradation of the male Y chromosome and determine the generation.

(See Dr. Nathaniel Jeanson's research with Answers in Genesis. This only works with a Young Earth viewpoint. And have a DNA sample taken, men, of your haplo group, then send it to Answers in Genesis to be added to the data—especially if you're Native American.)

Is this all in the Bible?

What we have in the Bible matches exactly with the model that I am putting forth, based on the vision given to me from God. Read the book and the Bible to determine for yourself if it's biblical. This is your responsibility.

9 | Model Summary & Q&A

Why didn't we already know this?

I don't know. I received the vision on 2/12/23.

Why did God give it to Gary Wentz?

I don't know. Maybe it's because I have never believed in anything but what the Bible says. Maybe God does have a sense of humor after all. Why else would He give it to someone who is:

- misunderstood more than most people in just casual conversation;
- very unimpressive, in all regards;
- not wealthy;
- dyslexic; hyperactive (*ADHD*); slightly autistic; depressed; anxious, awkward…

"And all of that is on top of being a white male, for crying out loud, and you know how bad those guys are these days. They can't get anything right!" … "Toxic white men."

Who else would have received a 4-second vision and turned it into a cosmology? I just don't really know why God gave this and the associated knowledge to me and not a million others. It's not like "I'm all that". And I don't know that I am the only one He has given it to. Maybe there are many more. (Since writing this, I have found another person who received a similar vision.)

What qualifies you to speak for God? Are you a prophet?

No, I'm not a prophet; I only received a vision from the Holy Spirit. It isn't so unusual. I will not consider myself to be qualified to speak for God until He tells me to, which hasn't happened yet. So I do not claim to speak for God. But though I do not speak for God, I do quote Him a lot. And I share what I think that He gave to me. He did not tell me to go and share what I have been sharing here. That is on me. That is why this is my cosmology model. I think that He wanted to leave a shadow of doubt in people's minds as to the origin of this idea. This model is my interpretation of the vision that God gave to me. I speak on my own. I believe that what I say is God inspired. Therefore, I have confidence in the model. But don't say that I speak for God—I don't claim that at all; I deny it. Also, prophecy is seeing the future; this is seeing the past.

Why do we need the Bible when we have Intelligent Design?

The relatively new descriptor, "Intelligent Design", brings a concept or view that attempts to marry two diametrically opposed concepts: Creation (*young earth*) and Deep Time (*old earth*). It is therefore an exercise in futility. It attempts to give God and His followers a more palatable compromise (*by at least acknowledging God's design*), while trying to do the same with the college crowd (*letting them have their Deep Time and Agnosticism too, as to Who actually did the designing and when*).

9 | Model Summary & Q&A

But the problems begin to arise instantly from the bottom up, like bubbles emerging in a pot of water starting to boil. This is because it is a false narrative that they attach to their "intelligent" arguments. If they wish to be intelligent about it, they will leave it the way God showed us in the Bible and build on it from there (*Biblical Creationism*). This is because wisdom begins with a healthy fear of, and reverence for, God. Therefore, I see Biblical Creationism as the best base for a scientific cosmology; especially when it fits the evidence the best, by far.

Acknowledging His design without acknowledging Him is a bit rude in my mind, if not just straight up insulting. In the same source where we hear of God (*the Bible*), we hear of a young Earth that He created not too long ago. Did God design the creation intelligently? Oh yeah, so much so that any attempt to take Him out of the picture, as some ID preachers attempt to do, is just foolish to the enlightened. "I will make foolish the intelligence of the intelligent." ~ God

If, by waving the flag of ID, instead of Biblical Creationism, you are denying the existence of the Creator, what have you gained in your reasoning (*fear of God is the beginning of wisdom*)? If, on the other hand, you are simply avoiding any connection to certain annoying people, then that is another issue. I would not blame you for avoiding those whom you find annoying. If you are not able to believe the words in the Bible, then that will be a problem for you, going forward. You are still in darkness if you are not getting your daily light from the Bible. If you do not understand what it says, read it in another version. If that doesn't work, ask someone or look it up. If that doesn't work, you need to really amp up your prayer about it. Hopefully, you have been praying all along anyway. Getting with others is a great way to figure it out; two heads are better than one, after all. But mostly (*firstly*), ask God and He will answer.

I see ID as just another bolt in the crossbow quiver. But use it wisely; it only goes so far. And you need to know how to use it. My advice is to stick to the Bible and incorporate it into the discussion of His Intelligent Design (*as I have done*).

Hasn't science proven evolution and deep time to be true, thereby disproving the Bible?

No, it hasn't. Actually, true science has completely debunked both evolution and deep time, showing them to be failed theories. (*DNA limits evolution; and stratification, one factor of many, buries deep time.*) And not only those failed concepts, but many of their related notions as well, like *string theory*, a *big bang*, and all of these *other dimensions* they want to introduce. We have four dimensions (*for now*) in our existence—TIME is the 4th. Anything else, whether fewer or more, is fantasy. We know that there is a Physical Realm and there is a Spiritual Realm, and can guess that God has/is His own Divine Realm beyond that. The failed notions of evil-you-shun and deep time go against all of the scientific evidence and solid logic pointing to those concepts (*as well as the Bible*). Evolution and Deep Time have not only been

9 | Model Summary & Q&A

unsupported by science; they have been solidly disproven by it; over and over.

And the related concepts that are attached to them are nullified as well, such as:

- string theory;
- five or more dimensions (*I do not consider spirit to be a dimension, but a different state or condition inside the 4 dimensions*);
- 1-, 2-, or 3-dimensional existence without all 4 together (*time is the 4th*);
- extraterrestrial races (*other than the angelic race—which includes our enemy*);
- a big bang (*without God doing it*)
- wormholes, and the like.

I don't know, I think it could be a long list. It's just crazy to me that people have been growing up and living their whole lives believing all or some of that stuff and some of them allowing it to take God out of their hearts and minds altogether.

And then there're those who think that they are so intelligent because they did not fall for the ancient stories that have been around for the entirety of human existence. How clever of them to not fall for someone's "piñata" cosmology, formed in the image of an alternate universe (*or multiverse*) with very little or no connection to this one.

The fact is, it is very easy to break the piñata. When the enemy presents God's story mixed with falsehood, it becomes a lie; a fabrication; a piñata. When the schools tell us that Noah's Flood (*for example*) happened a certain way (*other than how it actually happened*) and use their explanation as a straw man to beat up on, it becomes very easy to beat up on and rip to shreds, like a piñata. Continuing the example of the flood, there really is no way for a 40-day rainstorm, even if worldwide, to cover mountain peaks that soar ~29,000' above our current sea level. That is a piñata. There had to be more to the story—as my biblical model demonstrates. We can beat up the piñata a lot easier than we can beat up on the truth as given in God's word.

But their explanation or interpretation of what the Bible says isn't real; it's their own punching bag; a piñata that they constructed to beat up on. The real God Who made all of this (*and told us the story about it*) will not be amused when they see His face. They think that they are beating up on the Bible, when really they are just beating up on their own piñata that

9 | Model Summary & Q&A

they concocted in their minds. Doing so reinforces the false notions that fuel their fake knowledge and its fake power.

Making a mental piñata is actually very easy to do and is probably the basis of much delusion that we ALL feed to ourselves, or receive from the enemy. We all have done it and will continue to do it. We beat up on what we think is the truth, when really it was just another lie (*a piñata*) that we didn't want to fall for. Or in this case, we beat up on the truth, thinking it to be a lie, while embracing the real lies that come from the enemy. Hmm. Try not to be discouraged. This is why we all must walk in humility; because we are all judgmental toward others to some degree.

And as our scientific endeavors continually confirm that what the Bible says is true, we are at the same time nullifying false religions or other philosophies of men, with their false cosmologies and false gods (*not to mention the tyrannical chains they attach to peeps in the process*).

In fact, I would say that this revelation from God to me, and the ensuing instruction, has been showing me how all the world has been in a fog of falsity and fantasy ever since the generation after Tower of Babel incident. And even for the few to whom He has revealed the truth, they or their people have not been able to keep it pure and active (*I cite the Jews as the prime example—in brotherly love—since they did faithfully write it all down, even if they no longer believe it*).

Time, coupled with human frailty, has allowed the truth to be swept under the rug and walked on for ages, to the point that most folks (*nearly all*) are living in a dream world today. And although I have been given much insight into these things directly by God, I am seemingly doing all I can to forget all that He has taught me. I too am at least as frail in my fallen human state. It is only by God's gracious providence that I know any of this at all. And I am sure that it has been by His grace that I have not fallen victim to such outright lies as evil-you-shun and deep time (*evolution and an olde Earth, as taught in the schools*). Because I definitely embody the spirit of the quote given at the beginning of this essay:

> Remember, dear brothers and sisters, that few of you were wise in the world's eyes or powerful or wealthy when God called you. Instead, God chose things the world considers foolish in order to shame those who think they are wise. (1 Corinthians 1:26-27, NLT)

Be blessed as you seek the Lord, in reverent humility.

In Him and for you, not against you,

Gary

Other Books By Gary

End Times Made Easy

All you need to know about the End Times, without the usually unbiblical fluff or confusion. It's a solid look at what the Bible actually says about what is about to happen to this broken planet.

The Complete Gospel
— *a blending of the four Gospels into one continuous, flowing story*

Gary's first book puts the four Gospels, Matthew, Mark, Luke, and John, into a single continuous blending of the narrative of the life of Jesus Christ.

Rules For Life
According to Jesus

A collection of 33 commands given by our Lord and Savior on how to best live life, according to God's will and plan for the human family.

Gary The Gospel Guy Offline

A written collection of some of Gary's posts from his online blog. They talk of various aspects of life, death, salvation, and the human experience.

This Broken Planet

Visit Gary's weblog at:

 GaryTheGospelGuy.com

All of his books and more are available for no charge in web page format. That's right. You can read any or all of his complete books for free. You don't even have to give up your email address or name to do so. The website is also ad-free.

> "May the Lord bless you
> and protect you.
> May the Lord smile on you
> and be gracious to you.
> May the Lord show you His favor
> and give you His peace."
>
> (Numbers 6:24-26, NLT)